Boobytraps U.S. Army Instruction Manual Tactics, Techniques, and Skills

Plus

Operator's Manual for Machine Gun 7.62-mm, m60

Department of Defense

Pentagon
Publishing

CHAPTER 1

CHARACTERISTICS OF BOOBYTRAPS

Section I. INTRODUCTION

1. Purpose and Scope

a. This manual contains procedures, techniques, and expedients for the instruction of the soldier in the assembly, use, detection, and removal of boobytraps in combat.

b. Included are descriptions and discussions of the design and functioning characteristics of standard demolition items — firing devices, explosives, and accessories — and missiles, such as hand grenades, mortar ammunition, artillery ammunition, and bombs.

c. This manual also contains information on a variety of items and indigenous materials useful for improvising firing devices, explosives, and pyrotechnic mixtures for guerrilla warfare applications.

d. Factory-produced boobytraps (dirty trick devices) are described. Most of these have been developed and used in the field by foreign armies.

e. Safety measures pertinent to boobytrapping operations are provided for the protection of troops from casualty.

f. The contents of this manual are applicable to nuclear and non-nuclear warfare.

2. Comments

Users of this manual are encouraged to forward comments or recommendations for changes for improvement. Comments should be referenced to the page, paragraph, and line of text. The reason for each comment should be given to insure proper interpretation and evaluation. Forward all comments directly to the Commandant, U.S. Army Engineer School, Fort Belvoir, Virginia 22060.

Section II. PRINCIPLES OF OPERATION

3. Types of Boobytraps

A boobytrap is an explosive charge cunningly contrived to be fired by an unsuspecting person who disturbs an apparently harmless object or performs a presumably safe act. Two types are in use —improvised and manufactured. Improvised boobytraps are assembled from specially provided material or constructed from materials generally used for other purposes. Manfactured boobytraps are dirty trick devices made at a factory for issue to troops. They usually imitate some object or article that has souvenir appeal or that may be used by the target to advantage.

4. Assembling Boobytraps

A boobytrap consists of a main charge, firing device, standard base (not always used), and detonator. Another item, the universal destructor, is an adapter for installing a firing device assembly in a loaded projectile or bomb to make an improvised boobytrap. Also, firing device assemblies are often attached to the main charge by means of a length of detonating cord.

5. Boobytrap Firing Chain

THE FIRING CHAIN IS A SERIES OF INITIATIONS BEGINNING WITH A SMALL QUANTITY OF HIGHLY SENSITIVE EXPLOSIVE AND ENDING WITH A COMPARATIVELY LARGE QUANTITY OF INSENSITIVE EXPLOSIVE.

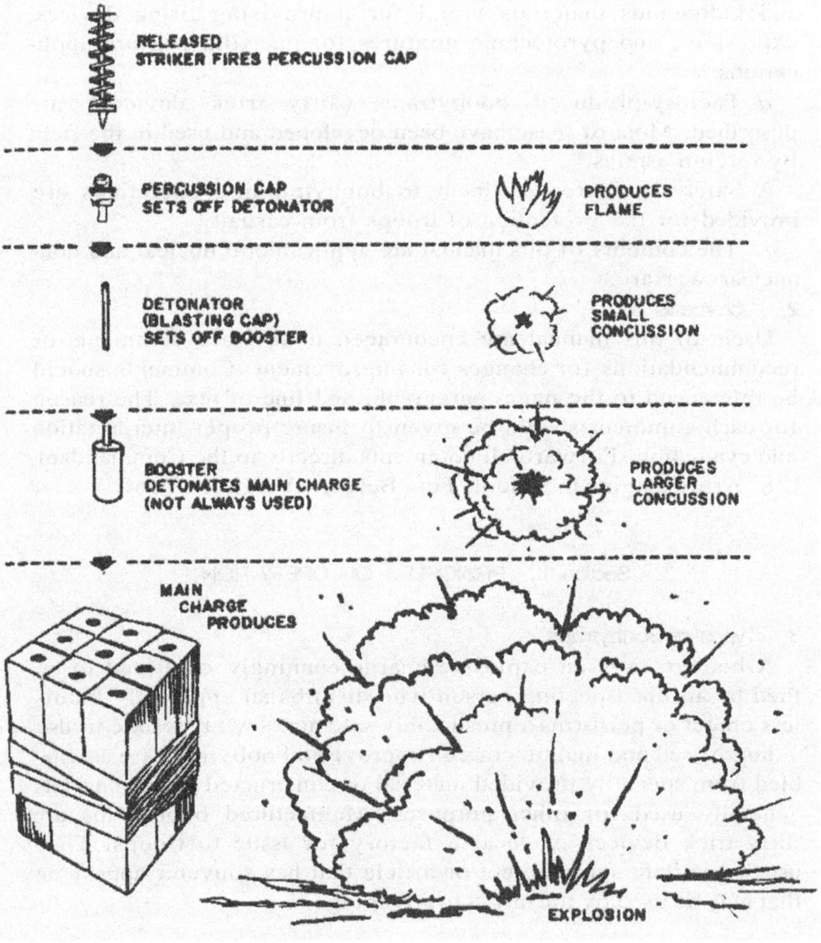

6. Initiating Actions

THE INITIATING ACTION STARTS THE SERIES OF EXPLOSIONS IN THE BOOBYTRAP FIRING CHAIN.

A. PRESSURE

WEIGHT OF FOOT STARTS EXPLOSIVE ACTION.

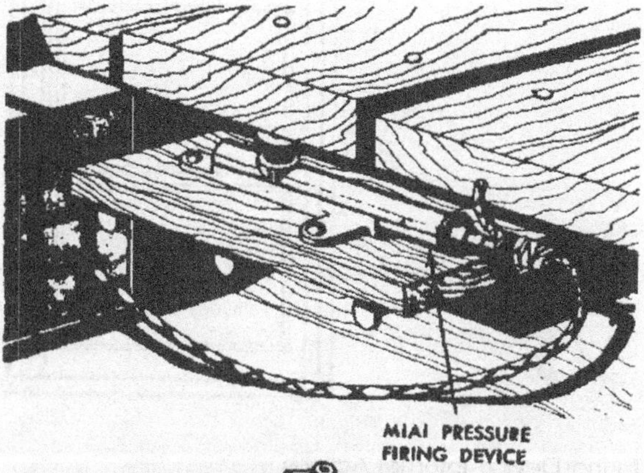

B. PULL

LIFTING THE SOUVENIR STARTS EXPLOSIVE ACTION.

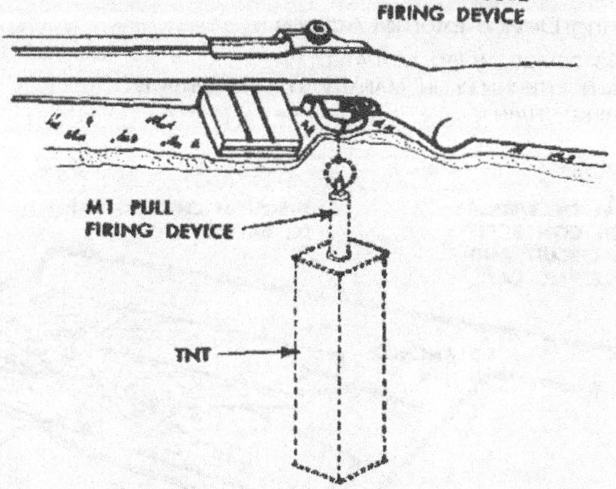

C. PRESSURE-RELEASE

MOVING THE STONE STARTS EXPLOSIVE ACTION.

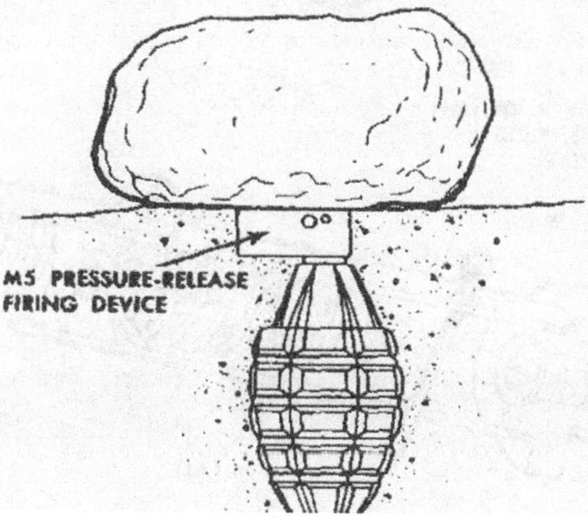

D. TENSION-RELEASE

RAISING LOWER SASH STARTS EXPLOSIVE ACTION.

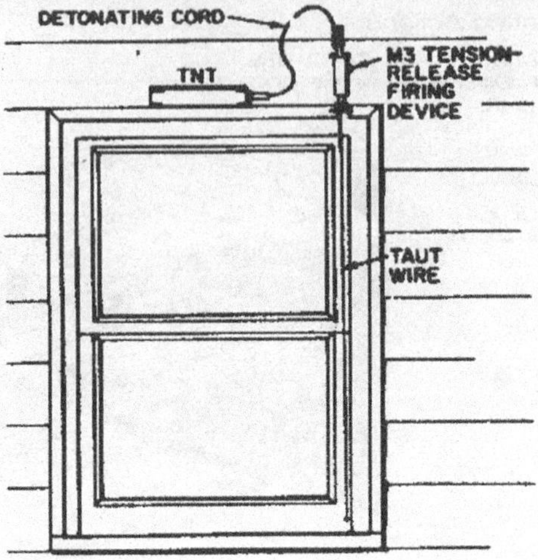

7. Firing Device Internal Actions

A FIRING DEVICE, WHEN ACTUATED MAY FUNCTION INTERNALLY IN MANY WAYS TO INITIATE THE FIRING CHAIN.

A. ELECTRIC

REMOVAL OF WEDGE BETWEEN CONTACTS CLOSES CIRCUIT AND FIRES ELECTRIC CAP.

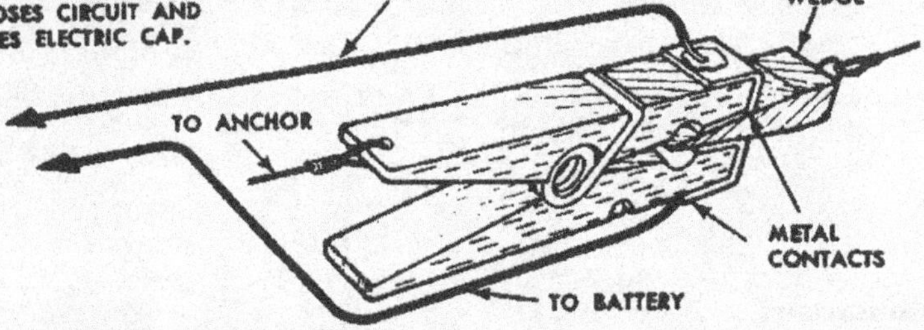

B. MECHANICAL

RELEASED STRIKER, DRIVEN BY ITS SPRING, FIRES PERCUSSION CAP.

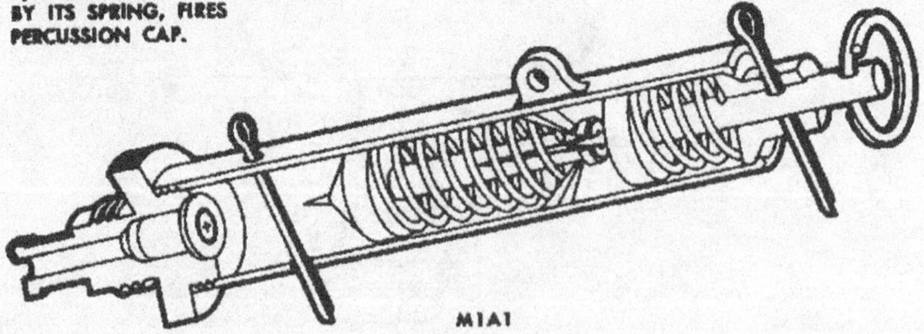

M1A1

C. PULL-FRICTION

PULLING THE CHEMICAL PELLET THROUGH THE CHEMICAL COMPOUND CAUSES FLASH THAT FIRES THE DETONATOR.

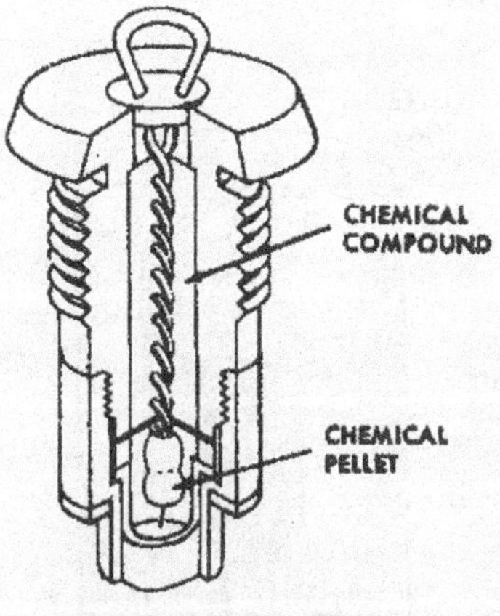

PULL-FRICTION FUZE, WEST
WW II GERMANY

D. PRESSURE-FRICTION

PRESSURE ON TOP OF THE STRIKER FORCES ITS CONE-SHAPED END INTO THE PHOSPHORUS AND GLASS MIXTURE IN THE MATING SLEEVE, CAUSING A FLASH THAT FIRES THE DETONATOR

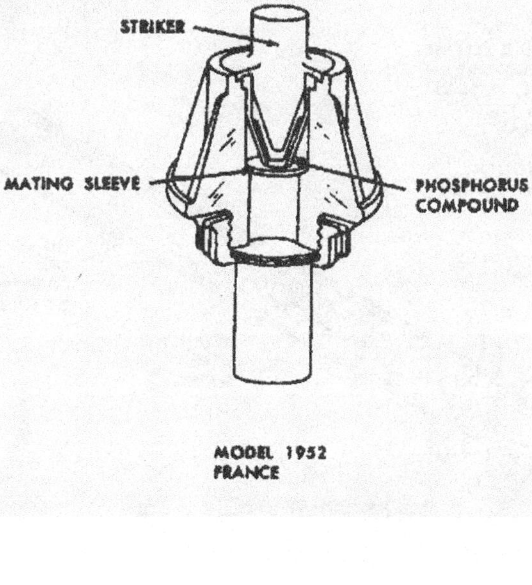

MODEL 1952
FRANCE

E. CHEMICAL

(1) PRESSURE

PRESSURE ON THE TOP BREAKS THE VIAL, FREEING THE SULPHURIC ACID TO MIX WITH THE FLASH POWDER, PRODUCING A FLAME THAT FIRES THE DETONATOR.

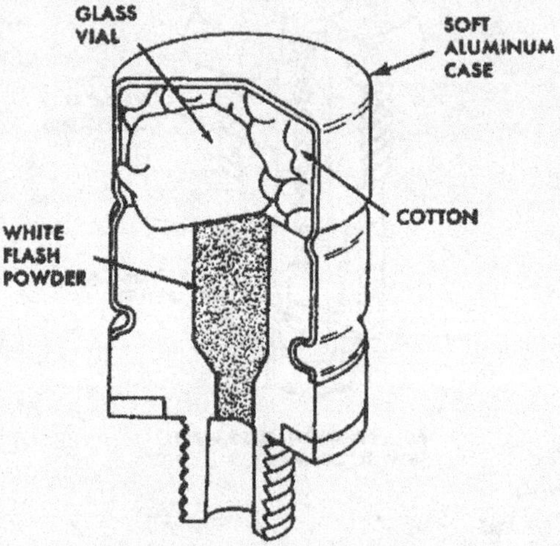

WW II GERMANY BUCK CHEMICAL FUZE

(2) DELAY

CRUSHING THE AMPULE RELEASES THE CHEMICAL TO CORRODE THE RETAINING WIRE, FREEING THE STRIKER TO FIRE THE DETONATOR. THE DELAY IS DETERMINED BY THE TIME NEEDED FOR THE CHEMICAL TO CORRODE THE RETAINING WIRE.

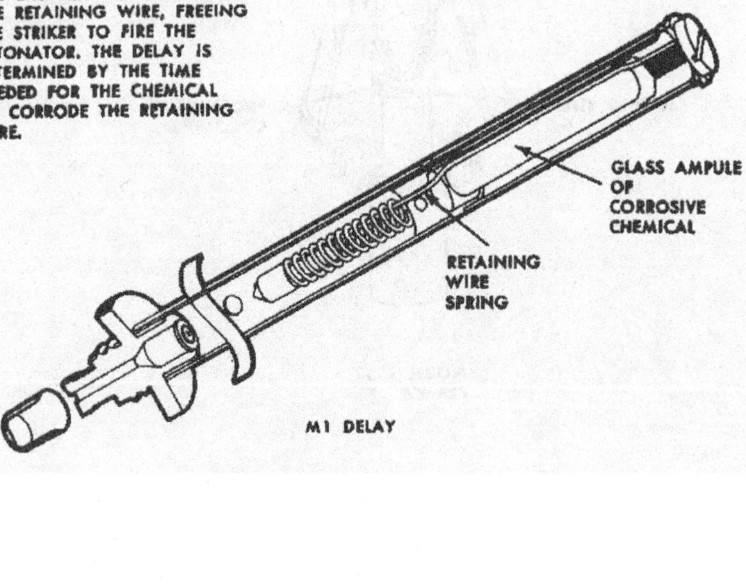

M1 DELAY

CHAPTER 2

USE OF BOOBYTRAPS

Section I. BASIC DOCTRINE

8. Tadkal Principles

Boobytraps supplement minefields by increasing their obstacle value. They add to the confusion of the enemy, inflict' casualties, destroy material, and lower morale. Boobytraps are usually laid by specialists. All military personnel, however, are trained in handL xplosives and other boobytrapping material, so that they may, ii necessary, boobytrap a mine or install a simple boobytrap.

9. Authority

a. Army commanders issue special instructions for the use of boobytraps within their command. Supplies are authorized and provided as required to meet boobytrapping needs.

b. Army and higher commanders may delegate authority to lay boobytraps to as low as division commanders. All higher commanders, however, may revoke this authority for a definite or indefinite period, as the tactical situation may require.

e. Records of all boobytraps laid are prepared and forwarded to higher headquarters.

d. Enemy boobytraped areas, as soon as discovered, are reported to higher headquarters to keep all interested troops advised of enemy activities. If possible, all boobytraps are neutralized; otherwise they are properly marked by warning signs.

Section II. PLANNING

10. Tadkal Effeds

a. The ingenious use of local resources and standard items is important in making effective boobytraps. They must be simple in construction, readily disguised, and deadly. They may produce unexpected results if conceived in sly cunning and built in various forms. Boobytraps cause uncertainty and suspicion in the mind of the enemy. They may surprise him, frustrate his plans, and inspire in his soldiers a fear of the unknown.

b. In withdrawal, boobytraps may be used in much the same way as nuisance mines. Buildings and other forms of shelter, roads, paths, diversions around obstacles, road blocks, bridges, fords, and similar areas are suitable locations for concealing boobytraps.

c. In defense, boobytraps, placed in the path of the enemy at strategic locations in sufficient numbers, may impede his progress, prevent detailed reconnoissance, and delay disarming and removal of minefields.

11. Basic Principles

Certain basic principles, as old as warfare itself, must be followed to get the optimum benefit from.boobytraps. Knowledge of these principles will aid the soldier, not only in placing boobytraps expertly, but in detecting and avoiding those of the enemy.

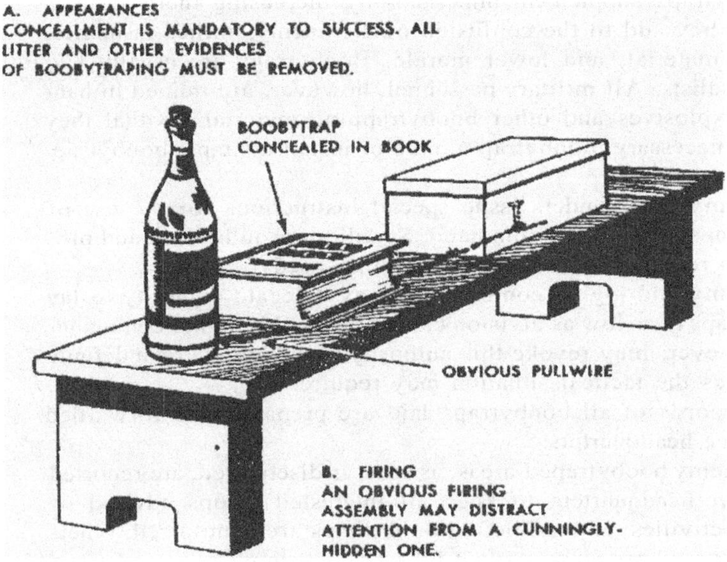

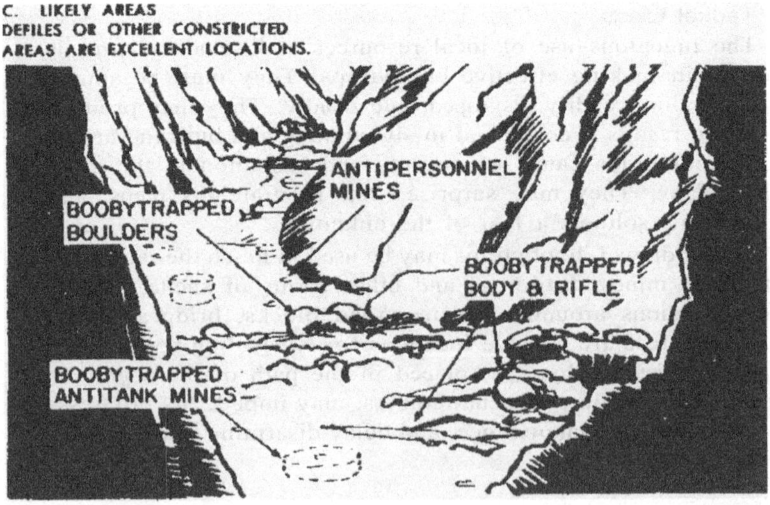

D. OBSTACLES
ROAD BLOCKS, FALLEN
TREES, LITTER, ETC.,
ARE IDEAL LOCATIONS

E. GATHERING PLACES
IN BUILDINGS, AT BUILDING ENTRANCES,
AND IN SIMILAR PLACES WHERE
SOLDIERS MAY MOVE OR
GATHER, DELAY CHARGES PAY OFF.

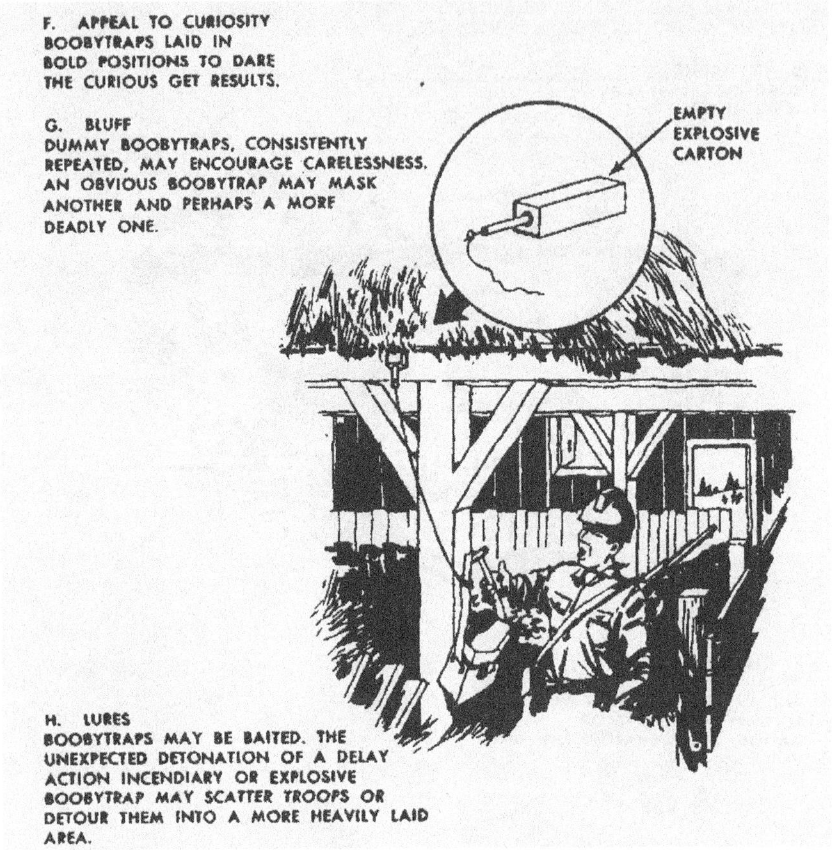

12. Location of Charges

a. Preparation. Small compact boobytraps are the most desireable for use in raids in enemy-held territory. Each member of a team must carry his own supplies and be able to operate independently. Boobytraps should be assembled, except for the attachment of the firing device, before entering enemy territory. This will reduce the work at the site to the minimum.

b. Location. Charges should be placed where they will do the most damage. A charge detonated against a stone wall will expend its force in magnified intensity away from the wall. The force of an explosion on the ground will affect the surrounding air more if the charge is placed on a hard surface. This deflects the explosive wave upward. A charge detonating 6 to 10 feet above the ground will damage a larger area than one laid on or below the surface.

c. Characteristics. Many inexpensive boobytraps, simple to make and easy to lay, will delay and confuse the enemy more than a small number of the expensive and complex kind. Complex mechanisms cost more, require more care in laying, and offer little more advantage than tiie simple type.

13. Reconnaissance

Complete reconnaissance of an area is essential to good planning. Without this and the preparation of a program, boobytraps may not be used effectively. Boobytrap teams are best suited to survey a combat area to determine its boobytrapping possibilities.

14. Plan of Operation

a. The commander with authority to use boobytraps coordinates his plans with other tactical plans. Timing of boobytrap operations with movement plans is extremely essential. Boobytraps should not be laid in areas where friendly troops will remain for any appreciable length of time. Plans will indicate what is to be done, where and when it will be done, and the troops to be used. Generally, trained troops are assigned such tasks.

b. The plan authorizes the use of boobytraps and the types and densities required in specified areas, depending on the terrain, time, personnel, and material available. The completion of the detailed plan is delegated to the commander responsible for installation. Materials are obtained from unit supply stocks on the basis of the proposed action.

c. Complete coordination between the troop commander and the officer supervising boobytrap activities is essential. The area should be evacuated immediately following the completion of the job.

d. The commander installing boobytraps prepares a detailed plan indicating the site and the location, number, type, and setting. He assigns boobytrap teams to specific areas and the laying of specified types. The plan covers arrangements for supplies and transportation and designates the location where all preliminary work on boobytraps will be done. Time tables are established to insure completion of the work to comply with withdrawal phases of tactical plans.

e. In hasty withdrawal, when there is no time for planning, each team will be given a supply of material with instructions for making the best possible use of it in the time allowed.

f. Boobytrap planning must give proper consideration to all known characteristics of the enemy. Members of teams should study the personal habits of enemy soldiers, constantly devising new methods to surprise them. Repetitions may soon become a pattern easily detected by an alert enemy.

g. Withdrawal operations are the most desirable of all for laying boobytraps. When an enemy meets a boobytrap at the first obstacle, his progress throughout the area will be delayed even though no others have been laid. A few deadly boobytraps and many dummies, laid indiscriminately, can inspire great caution. Dummies, however, should be unserviceable or useless items. Never throw away material that may return to plague friendly forces!

Section III. INSTALLATION

15. Responsibilities

a. A commander authorized to use boobytraps is responsible for all within his zone of command. He will keep adequate records showing their type, number, and location, and prepare information on those laid and on practices followed by the enemy.

b. Management of boobytrap services may be delegated to the engineer staff officer.

c. Unit commanders must know the location of all boobytraps in their areas and keep all subordinates so advised. Subordinates are also responsible for reporting to higher headquarters all new information obtained on enemy boobytraps.

d. Officers responsible for laying boobytraps prepare plans, supervise preliminary preparations, and direct their installation. They forward to proper authority a detailed report of their progress, advise all concerned when changes are made, and report to engineer intelligence units the discovery of any new enemy devices or low-cunning practices.

e. Engineer and infantry units, with special training, have the responsibility of installing and neutralizing boobytraps. Since adequate numbers of trainees may not always be available, all troops are given familiarity instruction in boobytrapping.

16. Procedures

Like all activities involving explosives, boobytrapping is dangerous only because of mistakes men make. Prescribed methods must be followed explicitly in the interest of personal safety and overall effectiveness.

a. Before assemblying a boobytrap, all components should be inspected for serviceability. They must be complete and in working order. All safeties and triggering devices must be checked to insure proper action, and for rust or dents that might interfere with mechanical action.

b. If a boobytrapping plan is not available, one must be prepared on arrival at the site, so that the material obtained will be required items only. A central control point should be established in each boobytrap area where supplies may be unloaded and from which directions may be given. In areas where many boobytraps are concentrated, safe passage routes from the control point to each location must be marked clearly. Lines or tape may be useful where vegetation is heavy. The control man is the key man.

c. Several teams may operate from one control point. Each team (rarely more than two men) is assigned to a specific area and supplies are issued only as needed. Each detail commander must make certain that every man knows his job and is competent to do it. Teams will remain separated so that one may not suffer from the mistake of another. When a job is completed, all teams must report to control man before going elsewhere.

d. One person in each team is designated leader to direct all work. If possible, members of a team will avoid working close together when a boobytrap is assembled. One member should do all technical work and the other be a helper to carry supplies, provide assistance needed, and learn the skills needed.

e. Boobytraps laid during raids into enemy held territory should be small, simple, and easily installed. Each member of a party must carry the supplies he needs. The use of boobytraps under these conditions, when accurate records are impossible, may be a hazard to friendly troops if raids into the same area should become necessary.

/. Procedure for installing boobytraps is as follows:

(1) Select the site that will produce the optimum effect when the boobytrap is actuated.

(2) Lay the charge, then protect and conceal it.

(3) Anchor the boobytrap securely, with nails, wire, rope, or wedges, if necessary.

(4) Camouflage or conceal, if necessary.

(5) Teams arm boobytraps systematically, working toward a safe area.

(6) Leave the boobytrapped area clean. Carry away all items that might betray the work that has been done, such as loose dirt, empty boxes, tape, and broken vegetation. Obliterate footprints.

17. Reporting, Recording, and Marking

Boobytraps are reported and recorded for the information of tactical commanders and the protection of friendly troops from casualty. Boobytrap installations are reported and recorded as nuisance minefields, whether the area contains both boobytraps and mines or boobytraps alone.

a. Reports

(1) *Intent.* This is transmitted *by* the fastest means available consistent with signal security. It includes the location of the boobytrapped area selected, the number and type of mines to be laid (if antitank mines are boobytrapped), boobytraps to be laid, the estimated starting and completing time, and the tactical purpose. The report is initiated by the commander authorized to lay the field and forwarded to higher headquarters.

(2) *Initiation of Laying.* This report is transmitted by the fastest means available consistent with signal security. It contains the location and extent of the field, total number of mines and boobytraps to be laid, and estimated time of completion. The commander of the unit installing

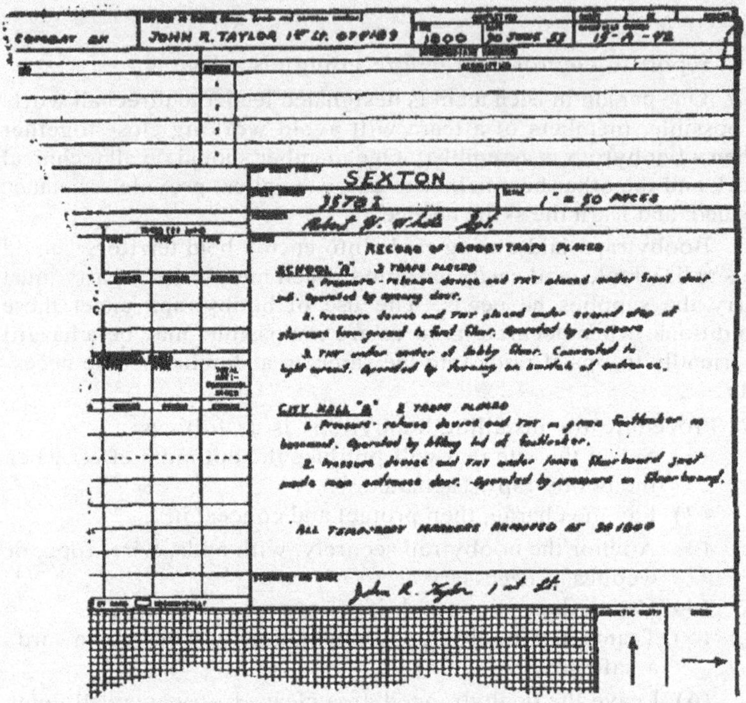

COMBAT BN | JOHN R. TAYLOR 1st LT. 077189 | 1800 | 30 JUNE 53 | 15-A-92

SEXTON

3570 I | 1" = 50 PACES

Robert W. White Sgt.

RECORD OF BOOBYTRAPS PLACED

SCHOOL "A" 3 TRAPS PLACED

1. Pressure-release device and TNT placed under front door sill. Operated by opening door.

2. Pressure device and TNT placed under second step of stairway from ground to first floor. Operated by pressure

3. TNT under center of lobby floor. Connected to light circuit. Operated by turning on switch at entrance.

CITY HALL "B" 2 TRAPS PLACED

1. Pressure-release device and TNT in green footlocker in basement. Operated by lifting lid of footlocker.

2. Pressure device and TNT under loose floorboard just inside main entrance door. Operated by pressure on floorboard.

ALL TEMPORARY MARKERS REMOVED AT 301800

John R. Taylor 1st Lt.

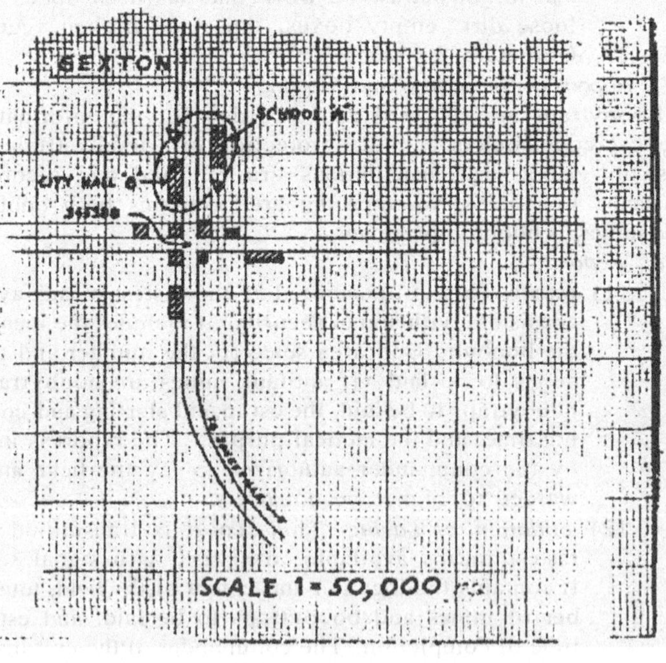

the field sends the report to the commander that directed him to lay it.

(3) *Completion.* The report of completion is transmitted by the fastest possible means. It contains the number and type of boobytraps laid, location and extent of the field or area and the time of completion. The report is forwarded to army level. When boobytraps are laid, either alone or with mines, the report of intent and the report of initiation of laying will include the estimated number of boobytraps to be placed and the report of completion, the number placed.

6. *Records.* Boobytraps are recorded as nuisance mine fields on the standard mine field record form. It is filled in as follows:

(1) The general locations are shown on the sketch, using the appropriate symbol. Boobytrapped areas or buildings are lettered serially, "A" being the nearest to the enemy.

(2) The number, types, locations, and methods of operation of boobytraps are entered in the NOTES section of the form. If space is lacking, additional sheets may be attached. If the boobytrap cannot be adequately described in a few short sentences, a sketch of minimum details will be included.

(3) The record is prepared simultaneously with the laying of the boobytrap and forwarded through channels to army level without delay. If a standard form is not available, the data required must be entered and submitted on an expedient form.

(4) Nuisance mine fields containing both mines and boobytraps are recorded as prescribed in FM 20-32. When the specific locations of boobytraps and manufactured devices cannot be accurately recorded (scattered laying in open areas) their number and type are entered in the notes section of the form and identified by grid coordinates.

e. Marking. Boobytraps are marked by special triangular signs painted red on both sides. On the side facing away from the danger area, a 3-inch diameter white disc, is centered in the triangle and the word BOOBYTRAPS is painted in white across the top in 1-inch letters. The STANAG or new sign is similar except for the 1-inch white stripe below the inscription. Signs may be made of metal, wood, plastic, or similar material. They are placed above ground, right-angled apex downwards, on wire fences, trees, or doors, windows, or other objects or by pushing the apex in the ground. These working signs are used by all troops to identify friendly boobytraps during the period preceding withdrawal from an area, or to warn friendly forces of the presence of active enemy boobytraps.

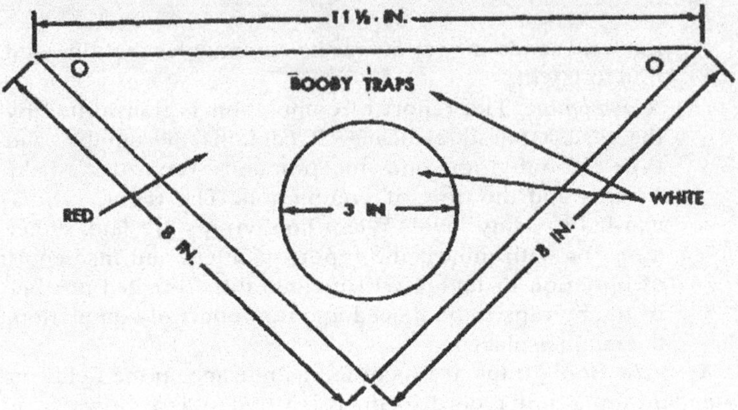

d. Abandonment. When abandoning a boobytrapped area to the enemy, all markers, wire, etc., are removed.

e. Signs. Signs are also used to mark enemy boobytraps or boobytrapped areas.

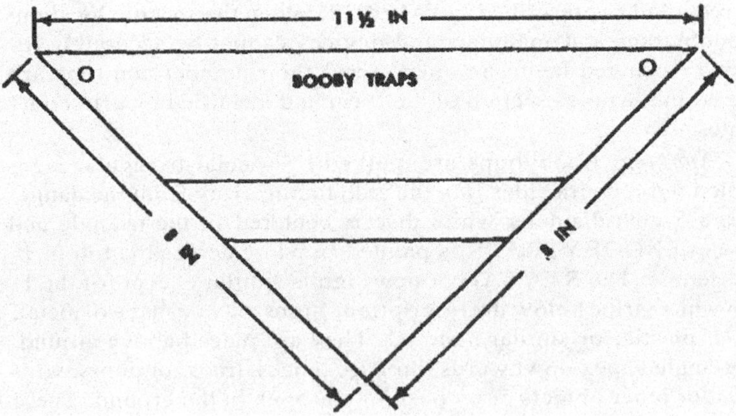

CHAPTER 3

BOOBYTRAPPING EQUIPMENT

Section I. FIRING DEVICES

18. Introduction

Many triggering devices are available for use in boobytraps. They include fuzes, igniters, and firing devices. All U.S. standard firing devices have the following advantages over improvisations; established supply, speed of installation, dependability of functioning, resistance to weather, and safety. All have a standard base coupling by which they may readily be attached to a variety of charges. For more detailed information see TM9-1375-200.

19. M1A1 Pressure Firing Device

a. Characteristics.

Case	Color	Dimensions D	Dimensions L	Internal Action	Initiating Action
Metal	OD	⅝ in	2¾ in	Spring-driven striker with keyhole slot release	20 lb pressure or more

Safeties	Accessories	Packaging
Safety clip and positive safety pin	3-pronged pressure head and extension rod	Five units with standard bases packed in cardboard carton. Thirty cartons shipped in wooden box.

b. Functioning.

A pressure of 20 pounds or more on the pressure cap moves the trigger pin downward until the striker spindle passes through the keyhole slot. This releases the striker to fire the percussion cap.

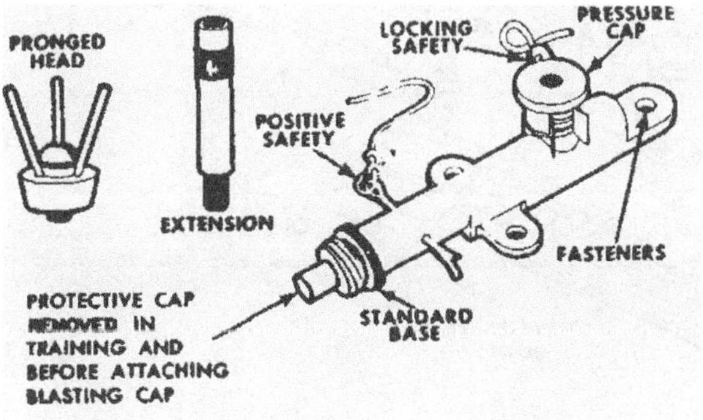

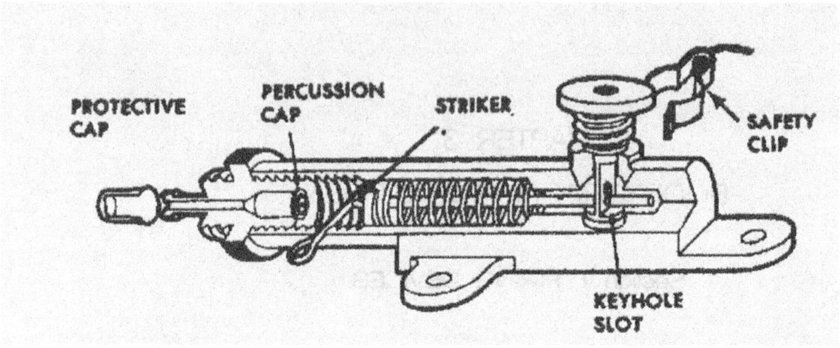

c. *Installing.*

(1) Remove protective cap from base and crimp on a non-electric blasting cap. *Crimper jaws should be placed* ***no*** *farther than # inch from open end of blasting cap.*

(2) Assemble 3-pronged pressure head and extension rod and screw in top of pressure cap, if needed.

(3) Attach firing device assembly to standard base.

(4) Attach firing device assembly to charge.

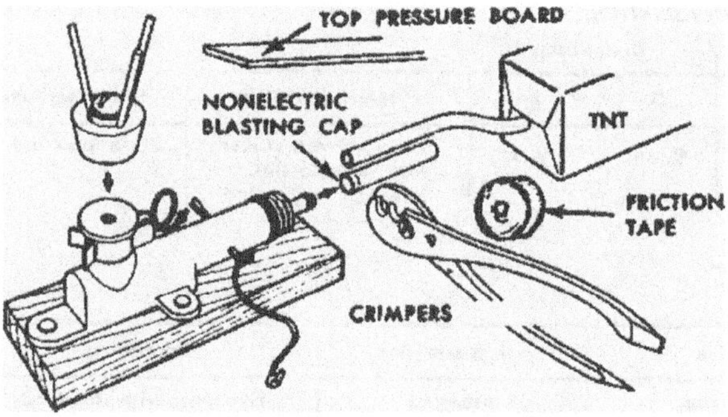

NOTE. If top pressure beard if used, allow clearance space between it and top of prongs or pressure cap.

d. Arming. Removs safety clip first and *positive pin last.*

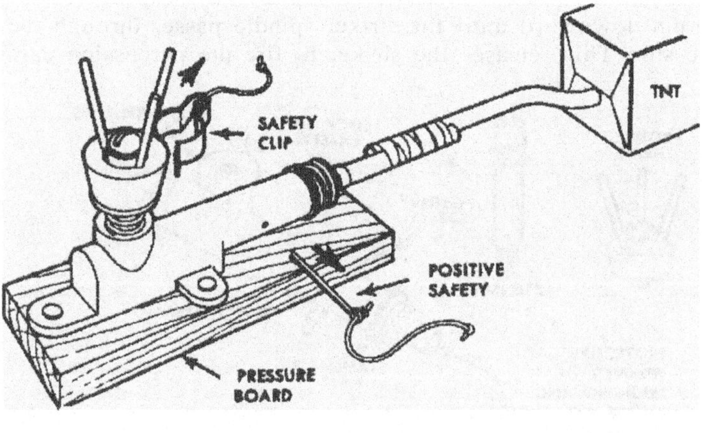

e. Disarming.

(1) Insert length of wire, nail, or original pin in positive safety pin hole.

(2) Replace safety clip, if available.

(3) Separate firing device and explosive block.

(3) Unscrew standard base assembly from firing device.

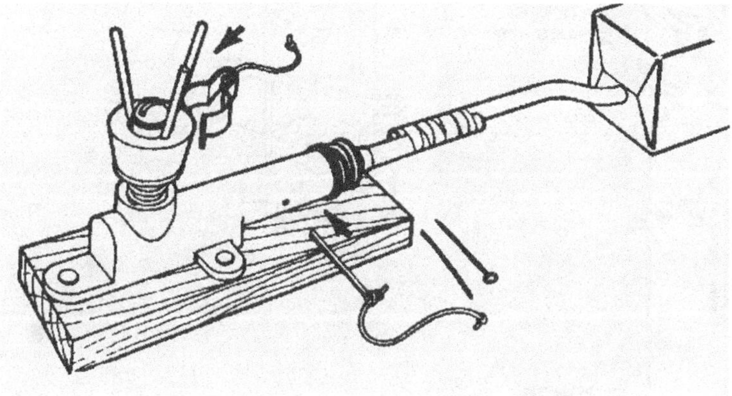

20. MI Pull Firing Device

a. *Characteristics.*

Case	Color	Dimensions D	Dimensions L	Internal Action	Initiating Action
Metal	OD	9/16 in	3 5/16 in	Mechanical with split-head striker release	3 to 5 lb pull on trip wire

Safeties	Packaging
Locking and positive safety pins	Five units complete with standard base and two 80-ft spools of trip wire, are packed in chipboard container. Thirty chipboard containers are packed in wooden box.

b. Functioning.

A pull of 3 to 5 lb. on trip wire .withdraws tapered end of release pin from split head of striker. This frees striker to fire the percussion cap.

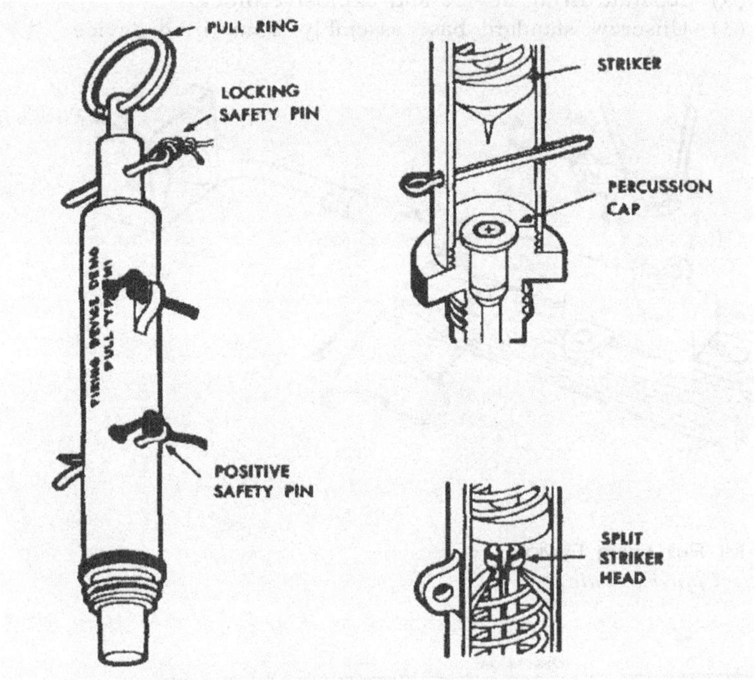

e. Installing.

(1) Remove protective cap.

(2) With crimpers, attach blasting cap to standard base. *Crimper jaws should be placed no farther than % in. from open end of blasting cap.*

(3) Attach firing device assembly to charge.

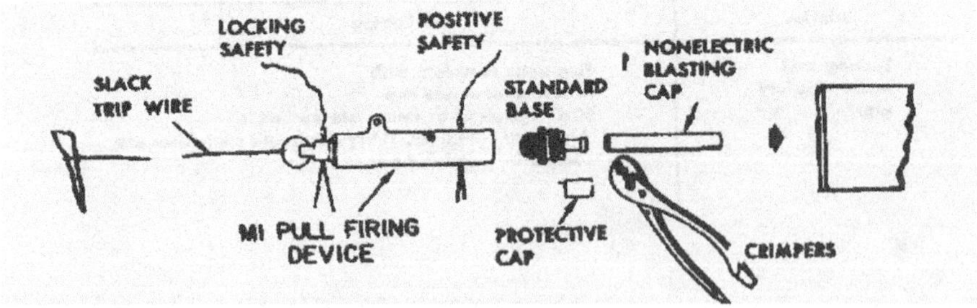

d. Arming.

(1) Anchor trip wire and fasten other end to pull ring.

(2) Remove locking safety pin first and *positive safety pin last.*

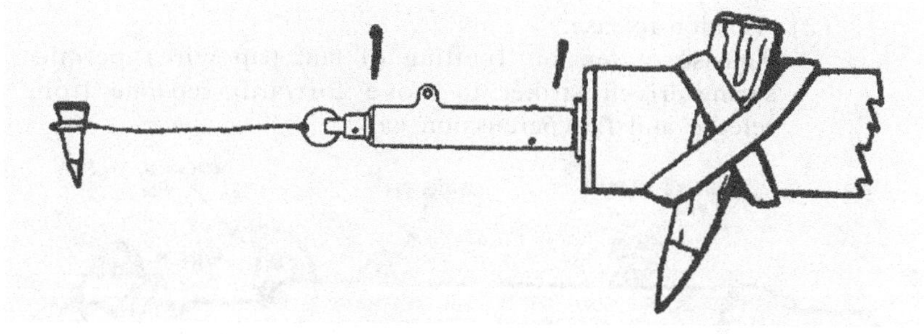

e. Disarming.

(1) Insert nail, length of wire, or original safety pin in positive safety pin hole *first.*

(2) Insert a similar pin in locking safety pin hole.

(3) Cut trip wire.

(4) Separate firing device and charge.

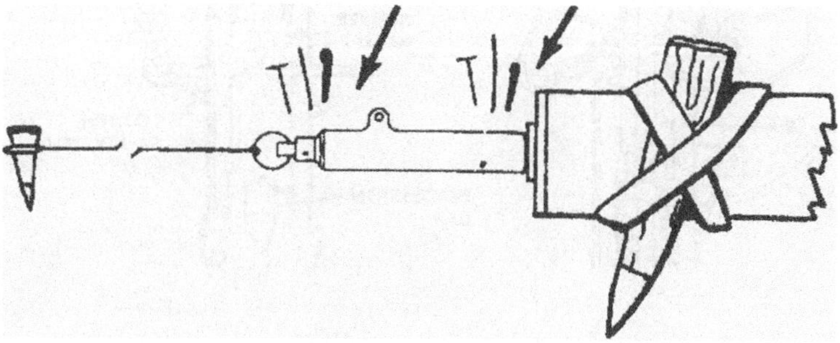

21. M3 Pull-Release Firing Device

a. Characteristics.

Case	Color	Dimensions D	Dimensions L	Internal Action	Initiating Action
Metal	OD	9/16 in	4 in	Mechanical with spreading striker head release	Direct pull of 6 to 10 lb or release of tension

Safeties	Packaging
Locking and positive safety pins	Five units with two 80-ft spools of trip wire in carton, and 5 cartons packed in wooden box

b. Functioning.

(1) Pull.

A pull of 6 to 10 lb. on taut trip wire raises release pin until shoulder passes constriction in barrel. The striker jaws then spring open, releasing striker to fire percussion cap.

(2) Tension-release.

Release of tension (cutting of taut trip wire) permits spring-driven striker to move forward, separate from release and fire percussion cap.

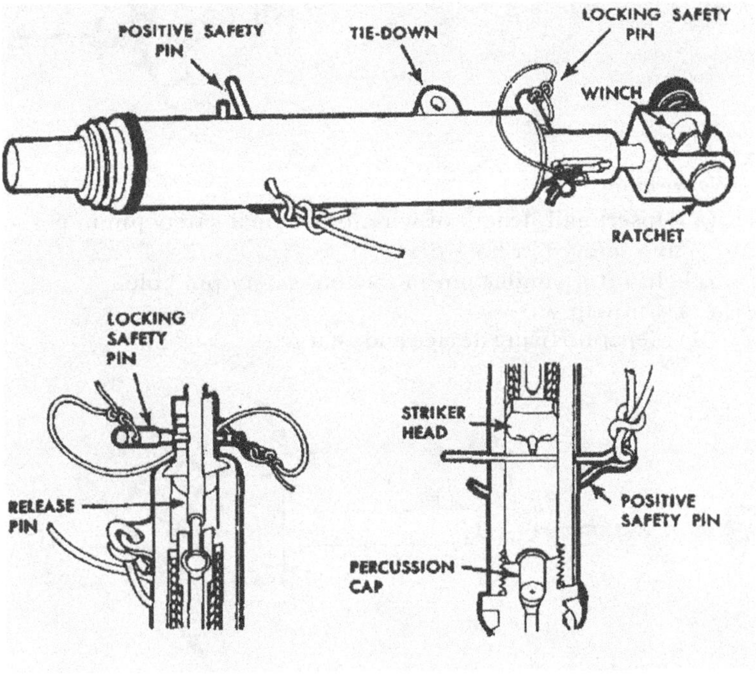

c. Installing.

(1) Remove protective cap.

(2) With crimpers, attach blasting cap to standard base. *Crimper jaws should be placed no farther than 1/4 in. from open end of blasting cap.*

(3) Attach firing device assembly to anchored charge (must be firm enough to withstand pull of at least 20 lb.).

(4) Secure one end of trip wire to anchor and place other end in hole in winch.

(5) With knurled knob draw up trip wire until locking safety pin is pulled into wide portion of safety pin hole.

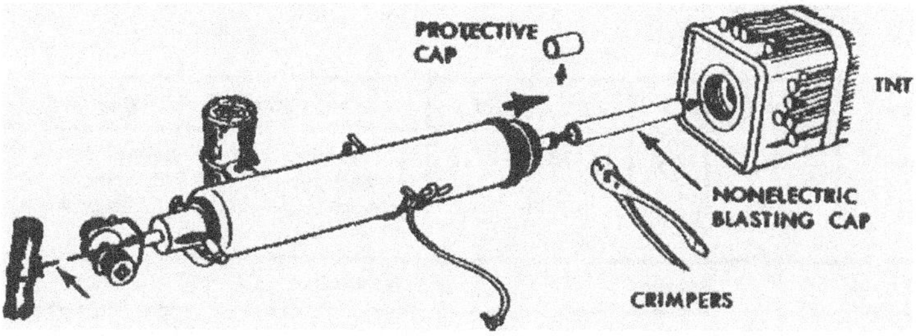

d. Arming.

(1) With cord, remove small cotter pin from locking safety pin and withdraw locking safety pin. If it does not pull out easily, adjust winch winding.

(2) With cord, pull out positive safety pin. This should pull out easily. If not, disassemble and inspect.

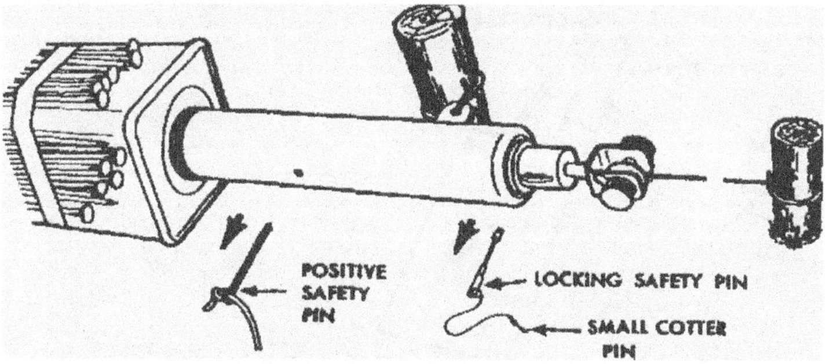

e. Disarming.

(1) Insert length of wire, nail, or cotter pin in positive safety pin hole.

(2) Insert length of wire, nail, of safety pin in locking safety pin hole.

(3) Check both ends and cut trip wire.

(4) Separate firing device from charge.

Note. Insert positive safety pin first. Cut trip wire last.

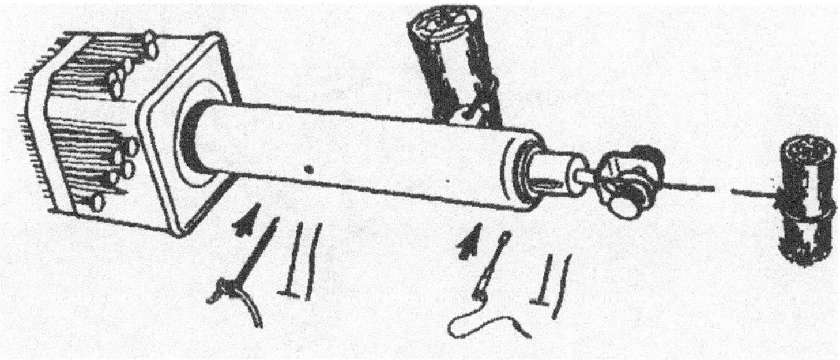

22. M5 Preiwre Releose Rring Device

a. Characteristics.

Case	Color	Dimensions L	W	H,T	Internal Action	Initiating Action
Metal	OD	1⅝	15/16	11/16	Mechanical with hinged plate release	removal of restraining wt, 5 lb or more

Accessories	Safeties	Packaging
Pressure board	Safety pin and hole for interceptor pin	Four firing devices complete and four plywood pressure boards in paper carton. Five cartons are packaged in fiber board box and 10 of these shipped in wooden box.

b. Functioning.

Lifting or removing retaining weight releases striker to fire the percussion cap.

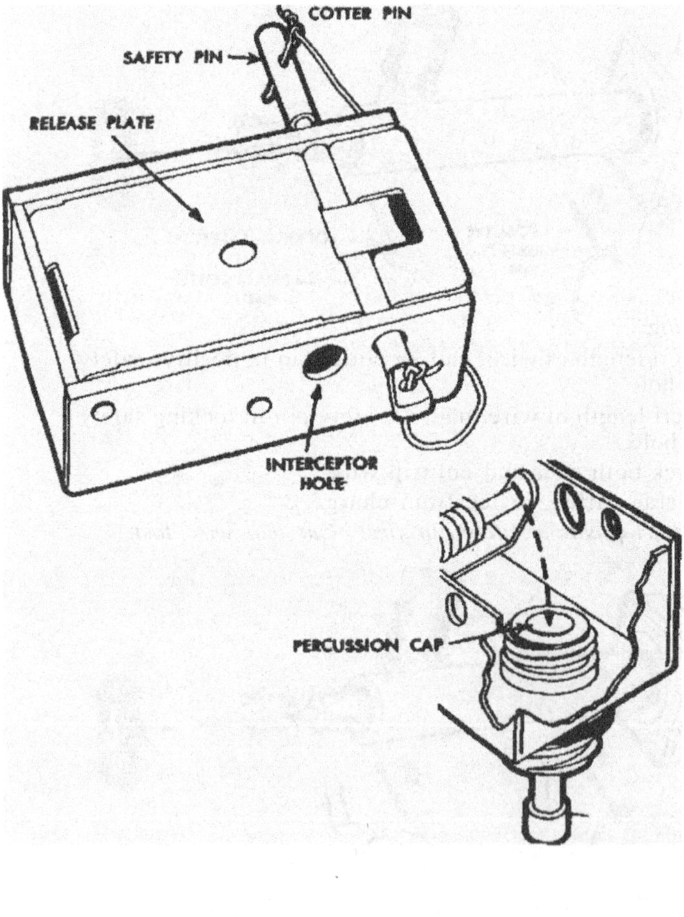

c. Installing.

(1) Insert a length of 10-gage wire in interceptor hole. Bend slightly to prevent dropping out.

(2) Remove small cotter pin from safety pin.

(3) Holding release plate down, replace safety pin with length of No. 18 wire. Bend wire slightly to prevent dropping out.

(4) Remove protective cap from base and with crimpers, attach blasting cap. *Crimper jaws should be placed no farther than ¹/i inch from open end of blasting cap.*

(5) Secure firing device assembly in charge.

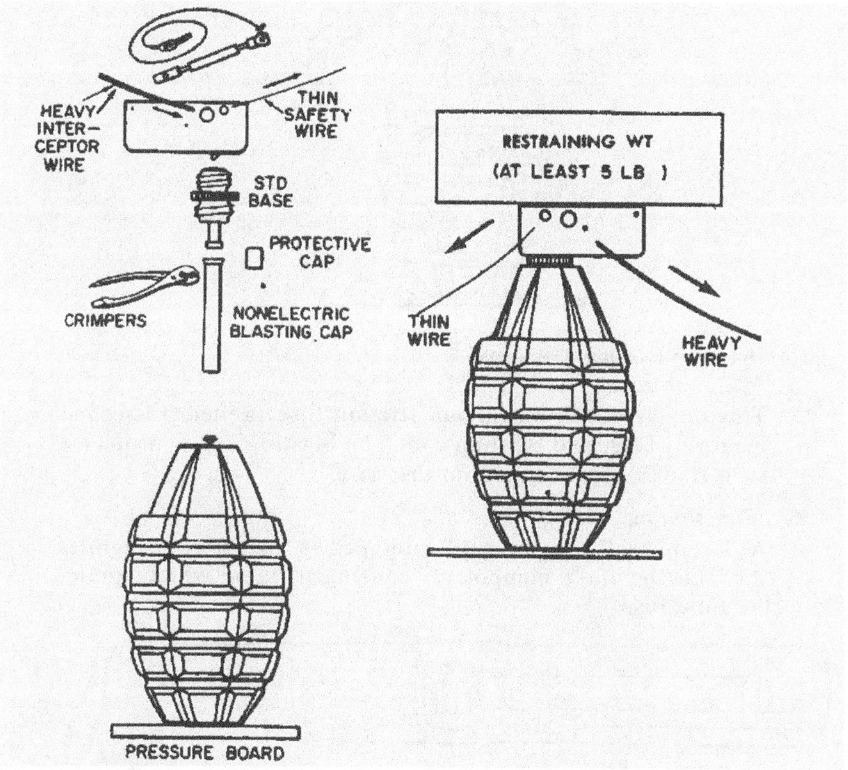

d. Arming.

(1) Place restraining weight on top of firing device.

(2) Remove thin wire from safety pin hole. If wire does not come out easily, restraining weight is either insufficient or improperly placed.

(3) Remove heavy wire from interceptor hole. It should move freely. *Note. Withdraw thin wire first and heavy wire last. Follow arming procedure carefully*

e. Disarming.

(1) Insert length of heavy gage wire in interceptor hole. Bend wire to prevent .dropping out. *Proceed carefully, as the slightest disturbance of the restraining weight might initiate the firing device.*

(2) Separate firing device from charge.

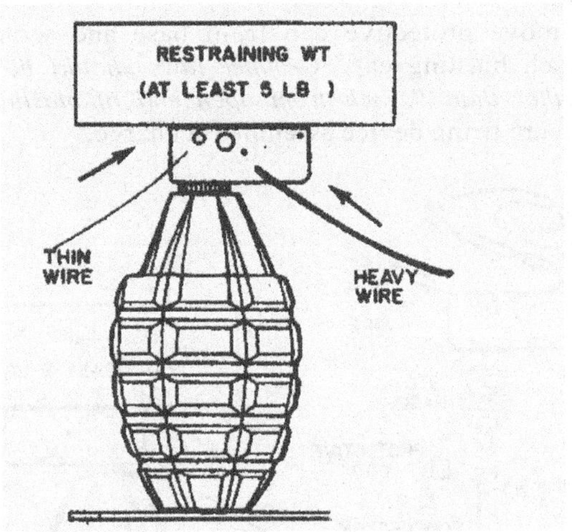

23. 15-Second Delay Detonator

a. *Characteristics.*

This device consists of a pull-friction fuse igniter, 15-second length of fuse, and blasting cap. The blasting cap is protected by a transit cap screwed on the base.

6. *Functioning.*

A strong pull on the pull ring draws the friction igniter through the flash compound, causing a flame which ignites the time fuse.

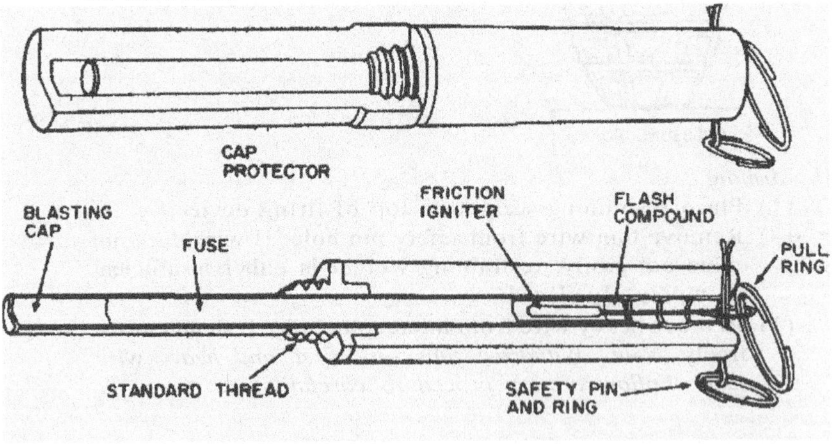

c. Installing.

(1) Unscrew transit cap from base.

(2) Secure device in charge.

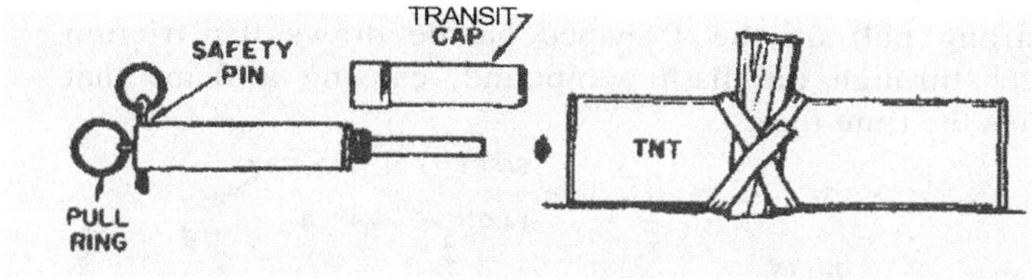

d. Arming.

(1) *Manual initiation.* Remove safety pin.

(2) *Trip wire initiation.*

(a) Attach one end of trip wire to anchor stake and the other to pull ring.

(b) Remove safety pin.

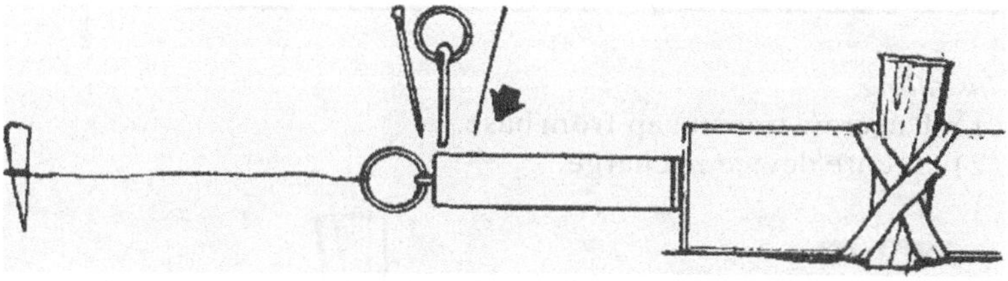

e. Disarming.

(1) Insert length of wire, nail, or original safety pin in safety pin hole.

(2) Remove trip wire.

(3) Separate firing device from charge.

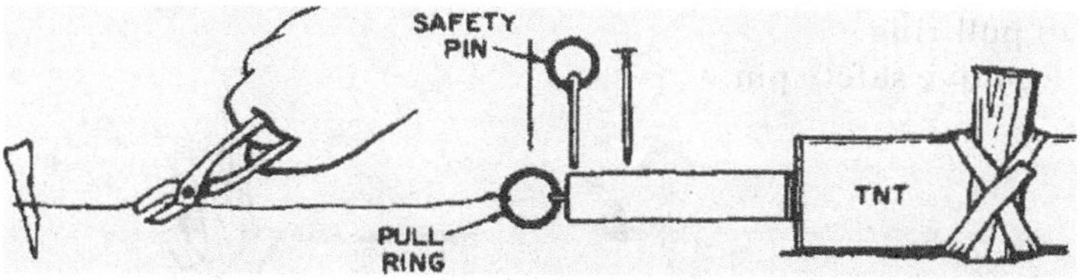

24. 8-Second Delay Detonator

a. Characteristics.

This device consists of a pull-type fuse lighter, 8-second length of fuse, and a blasting cap. The blasting cap is protected by a transit cap, screwed on the base.

b. Functioning.

A strong pull on the T-shaped handle draws the friction igniter through the flash compound, causing a flame that ignites the time fuse.

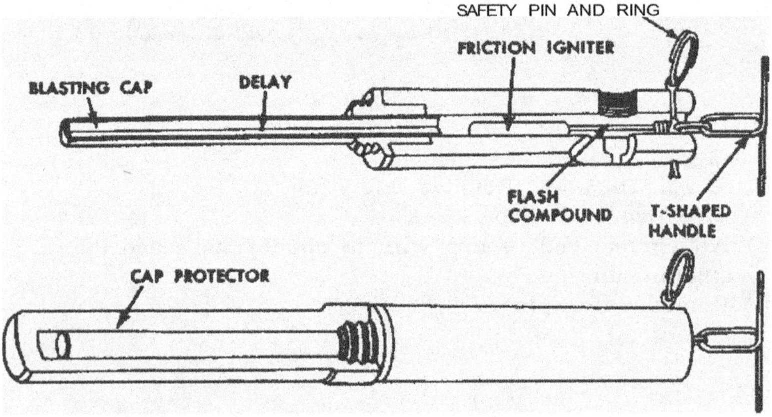

c. Installing.

(1) Unscrew transit cap from base.

(2) Secure device in charge.

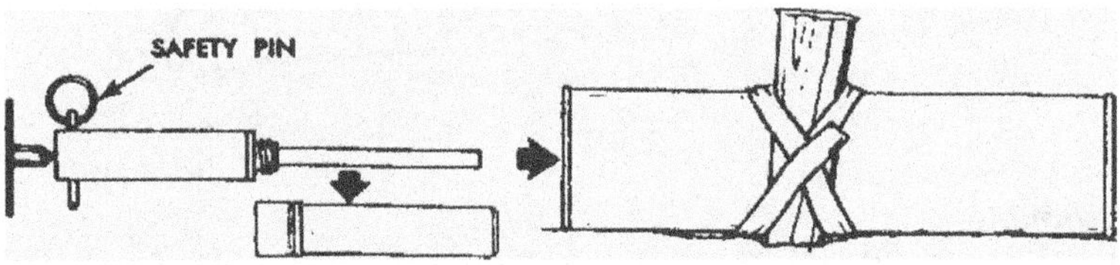

d. Arming.

(1) Manual initiation: Remove safety pin.

(2) Trip wire initiation.

(a) Attach one end of trip wire to anchor stake and the other to pull ring.

(b) Remove safety pin.

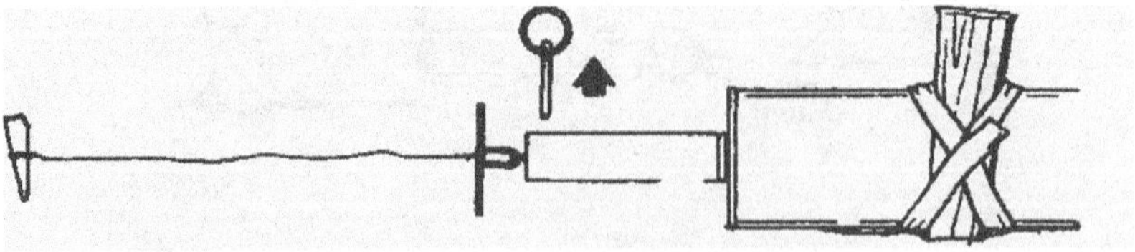

e. Disarming.

(1) Insert length of wire, nail, or safety pin in safety pin hole.

(2) Remove trip wire.

(3) Separate firing device from charge.

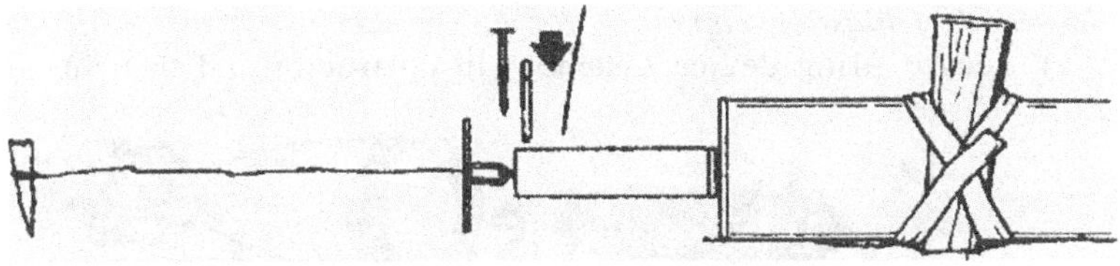

25. MI Delay Firing Device

a. Characteristics.

Case	Color	Dimensions D	Dimensions L	Internal Action	Delay
Copper and brass	Natural Metal	7/16 in	6¼ in	Mechanical with corrosive chemical release	4 min to 9 dos, identified by color of safety strip

Safety	Packaging
Colored strip inserted in hole above percussion cap.	10 units—2 red, 3 white, 3 green, 1 yellow, and 1 blue—and a time delay temperature chart packed in paper board carton, 10 cartons in fiber board box, and 5 boxes in wooden box.

b. Functioning.

Squeezing copper half of case crushes ampule, releasing chemical to corrode restraining wire and release striker.

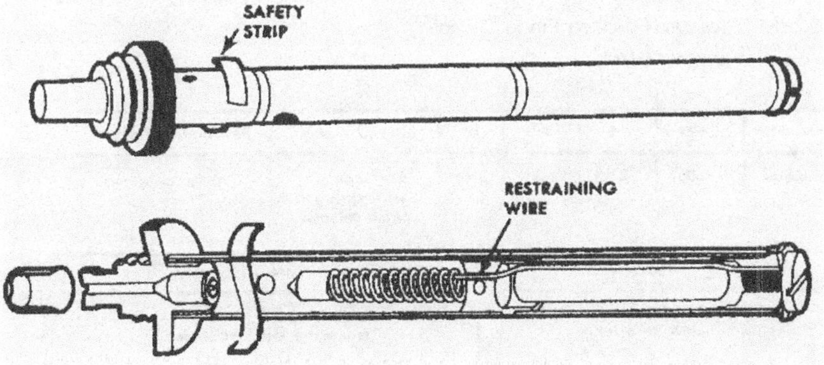

c. *Installing.*

(1) Select device of proper delay.

(2) Insert nail in inspection hole to make sure that firing pin has not been released.

(3) Remove protective cap from base.

(4) With crimpers, attach blasting cap to base. *Crimper jaws should be placed no farther than 1/4 in. from open end of blasting cap.*

(5) Secure firing device assembly in destructor and then in charge.

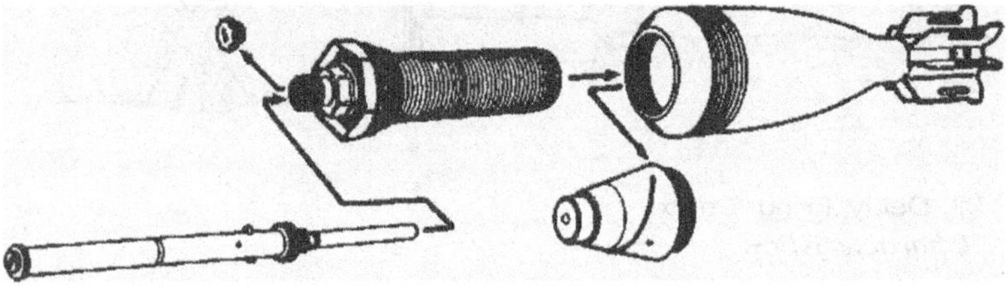

d. *Arming.*

(1) Crush ampule by squeezing the copper portion of case.

(2) Remove safety strip.

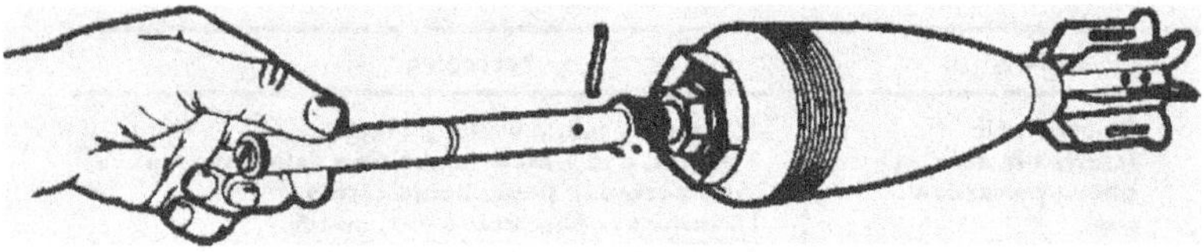

e. *Disarming.*

There is no safe way of disarming this firing device. If disarming is necessary, insert an improvised safety pin through inspection holes.

26. MI Prcssuro-Raioasa Firing Devics

a. *Characteristics.*

Case	Color	Dimensions L	W	Ht	Internal Action	Restraining Pressure
Metal	OD	3 in	2 in	2 in	Mechanical with springed latch release	3 lb or more

Safeties	Issue
Safety pin and hole for interceptor pin	Obsolete, but many are still available

6. *Functioning.*
Lifting or removing restraining weight unlatches lever, releasing striker to fire percussion cap.

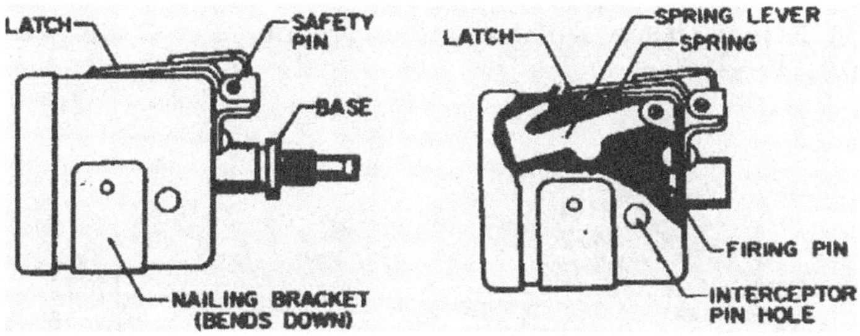

c. *Installing.*
(1) Insert a length of heavy gage wire in interceptor hole. Bend slightly to prevent dropping out.
(2) Holding down latch, remove safety pin and replace with length of thin wire.
(3) Remove protective cap from base and with crimpers attach nonelectric blasting cap. *Crimper jaws should be placed no farther than % in. from open end of blasting cap.*
(4) Assemble length of detonating cord, priming adapter, nonelectric blasting cap, and explosive block.
(5) Attach free end of detonating cord to blasting cap on Ml release device with friction tape, allowing 6 in. of detonating cord to extend beyond joint.

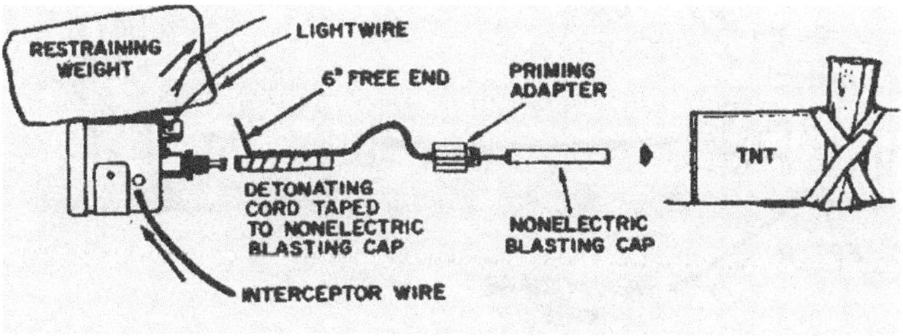

d. Arming.

(1) Place restraining weight on top of tiring device.

(2) Remove thin wire from safety pin hole. If it does not come out easily, restraining weight is either insufficient or improperly placed.

(3) Remove heavy wire from interceptor hole.

Note. Proceed carefully.

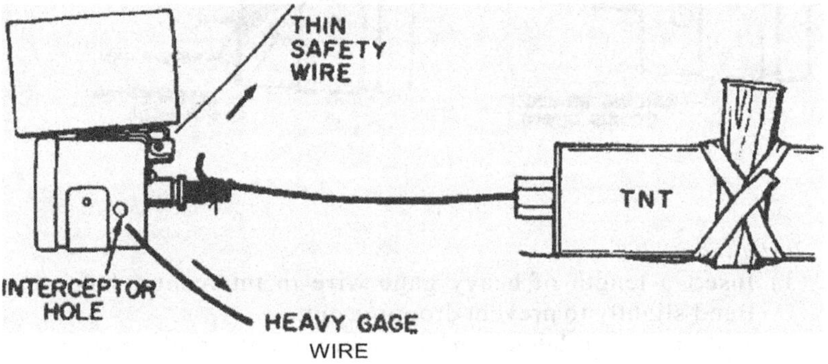

e. Disarming.

(1) *Proceed carefully as the slightest disturbance of restraining weight might unlatch lever and detonate the mine.* Insert length of heavy gage wire in interceptor hole. Bend wire to prevent dropping out.,

(2) Insert length of thin wire in safety pin hole, if possibile.

(3) Separate firing device assembly and explosive charge.

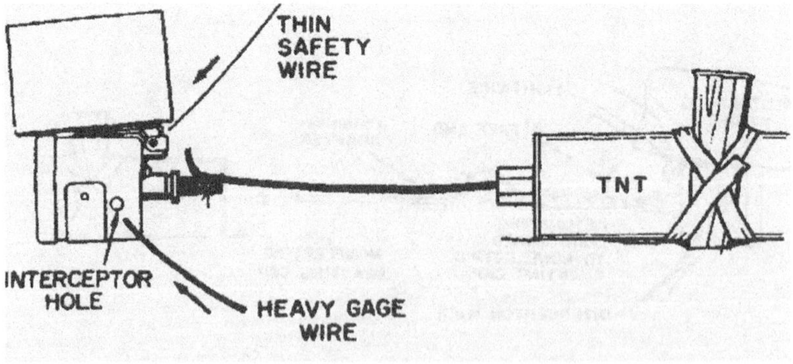

27. Explosives and 'Accessories (For more detailed information, see FM5-25 and TM 9-1375-200.)

a. TNT. This is issued in **14,** 1/2 and 1-pound blocks in a cardboard container with lacquered metal ends. One end has a threaded cap well. Half-pound blocks are obtained by cutting a 1-pound package in the center.

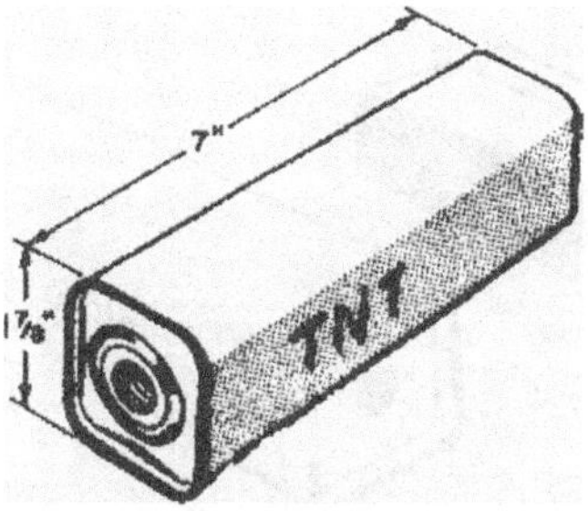

b. Ml Chain Demolition Blocks (Tetrytol). This explosive consists of eight 21/^-pound tetrytol blocks cast 8 inches apart onto a single line of detonating cord, which extends 2 feet beyond the end blocks. All blocks have a tetryl booster in each end. Each chain is packed in a haversack, and two haversacks in a wooden box.

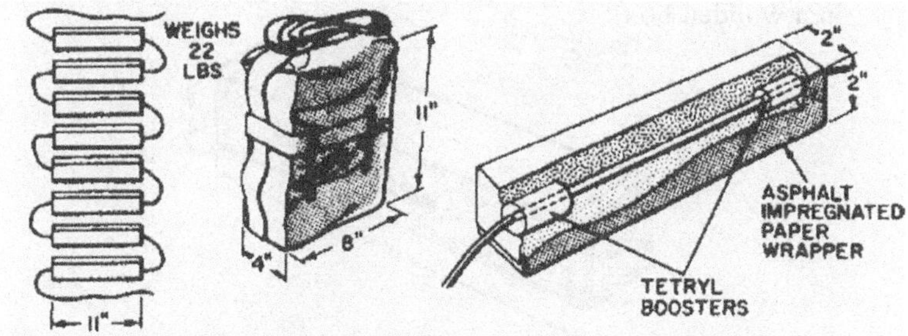

c. M2 Demolition Block (Tetrytol). The M2 demolition block i enclosed in an asphalt impregnated paper wrapper. It has ; threaded cap well in each end. Eight blocks are packed in ; haversack, and two haversacks in a wooden box.

d. *MS and MS Demolition Blocks (Composition C3).* These consist of a yellow, odorous, plastic explosive more powerful than TNT. The M3 block has a cardboard wrapper perforated around the middle for easy opening. The M5 block has a plastic container with a threaded cap well. Eight M3 or M5 blocks are packed in a haversack; and two haversacks, in a wooden box.

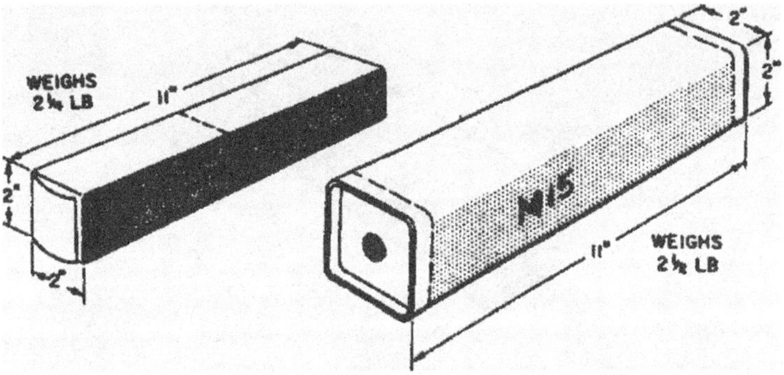

e. *M5Al Demolition Block (Composition CU).* This is a white plastic explosive more powerful than TNT, but without the odor of C3. Each block is wrapped in plastic covering with a threaded cap well in each end. Twenty-four blocks are packed in a wooden box.

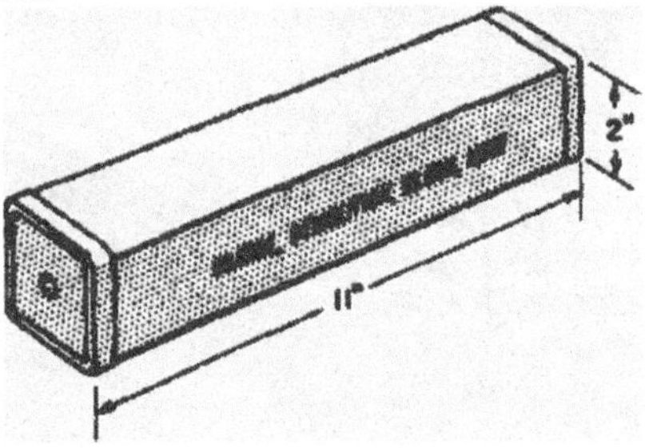

M112 Demolition Charge (Composition CU). This is composition C4 in a new package measuring 1 in. x 2 in. x 12 in. Each block has an adhesive compound on one face. Further information is not available.

g. *M118 Demolition Charge.* The M118 charge is composed of PETN and plasticizers. The detonating rate is approximately 23,000 ft. per second. Each package contains four sheets % in. x 3 in. x 12 in. Each sheet has an adhesive compound on one face. Further information is unavailable.

h. *Composition B.* Composition B is a high explosive with a relative effectiveness higher than TNT, and more sensitive.

Because of its high dentonation rate and shattering power, it is used in certain bangalore torpedoes and in shaped charges.

PETN. This is used in detonating cord. It is one of the most powerful military explosives, almost equal to nitroglycerine and RDX. In detonating cord, PETN has a velocity rate of 21,000 feet per second.

Amatol. Amatol, a mixture of ammonium nitrate and TNT, has a relative effectiveness higher than that of TNT. Amatol (80/20) is used in the bangalore torpedo.

k. *RDX.* This is the base charge in the M6 and M7 electric and nonelectric blasting caps. It is highly sensitive, and has a shattering effect second only to nitroglycerine.

. *Detonating Cord.*

(1) *Types I and II.* These consist of a flexible braided seamless cotton tube filled with PETN. On the outside is a layer of asphalt covered by a layer of rayon with a wax gum composition finish. Type II has the larger diameter and greater tensile strength.

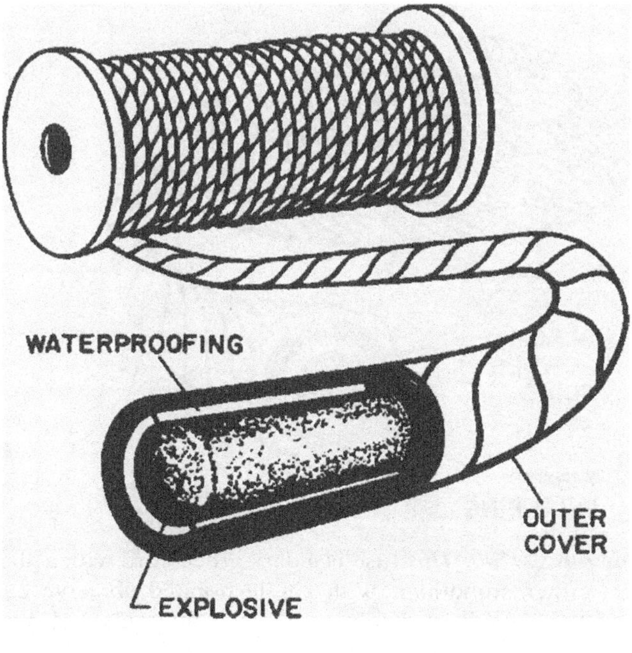

(2) ***Type IV*. This is similar to types I and II, except for the special smooth plastic covering designed for vigorous use and rough weather.**

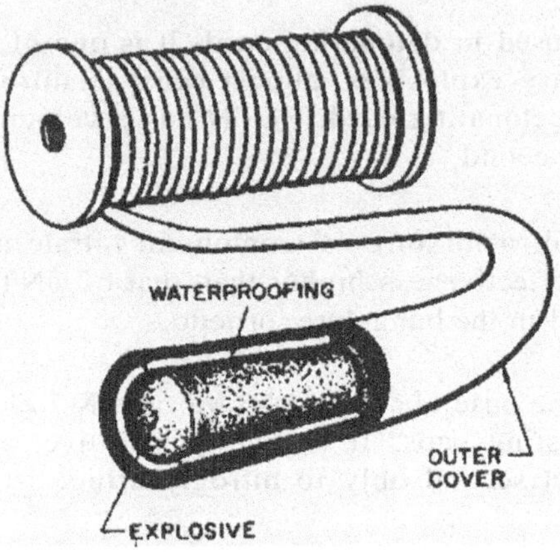

m. Blasting Time Fuse. This consists of black powder tightly wrapped in layers of fabric and waterproofing materials. It may be any color, orange being the most common. As burning rate varies from about 30 to 45 seconds per foot, each roll must be tested before using by burning and timing a 1-foot length.

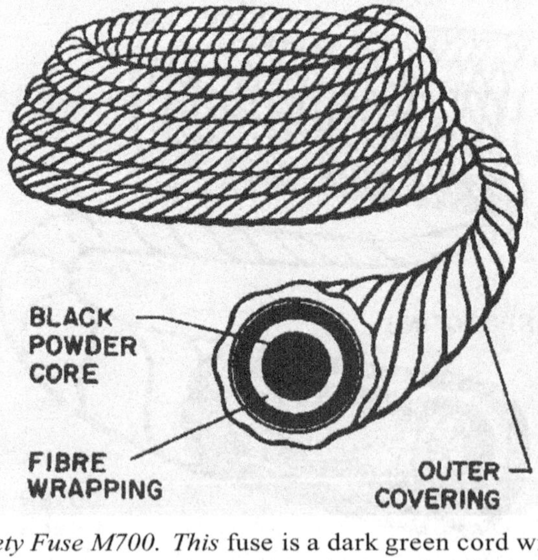

. *Safety Fuse M700. This* fuse is a dark green cord with a plastic cover, either smooth or with single pointed abrasive bands around the outside at 1-foot or 18-inch intervals and double painted abrasive bands at 5-foot or 90-inch intervals. Although the burning rate is uniform (about 40 seconds per foot), it should be tested before using by burning and timing a 1-foot length.

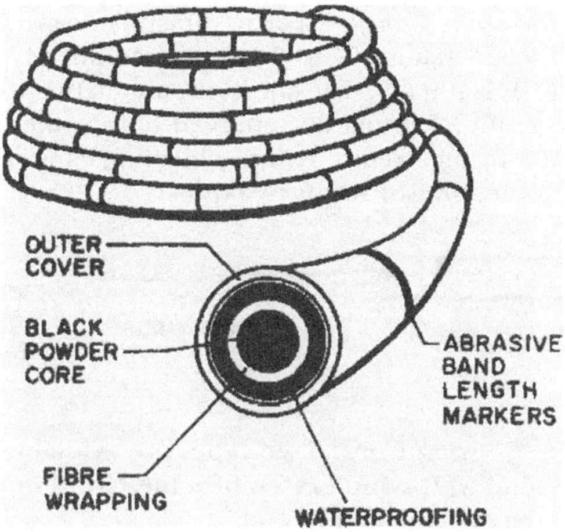

o. M60 Fuse Lighter.

(1) *To install; Unscrew the fuse holder cap, remove shipping* plug, insert time fuse, and tighten cap.

(2) *To reload:*

(a) Insert primer base and primer in end of lighter housing.

(b) Put washers and grommets in open end of fuse holder cap as shown, and screw fuse holder cap firmly on housing.

(c) Unscrew fuse holder cap about three turns and insert a freshly cut end of time fuse into the hole in the cap until it rests against the primer.

(d) Tighten cap.

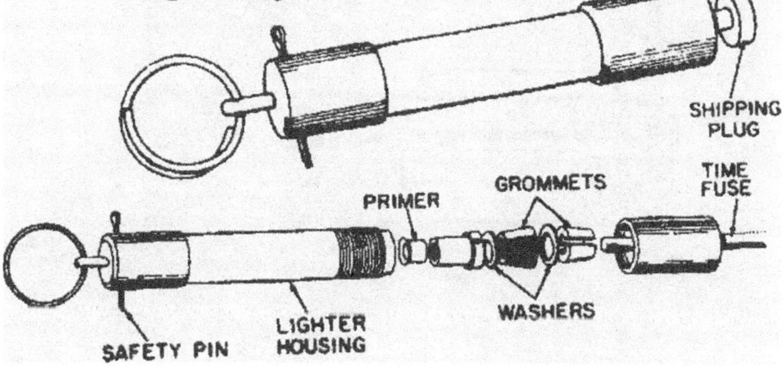

(3) *To fire:*

(a) Remove safety pin

(b) Pull on pull ring. '

Note. Lighter is reusable after the insertion of a new primer and the reassembly of parts.

p. Electric Blasting Caps. Electric blasting "caps have three lengths of leads—short (4 to 10 ft.), medium (12 to 14 ft), and long (50 to 100 ft). The short-circuit tab or shunt prevents accidental firing. It must be removed before the cap is connected in the firing circuit. Military blasting caps are required to insure detonation of military explosives.

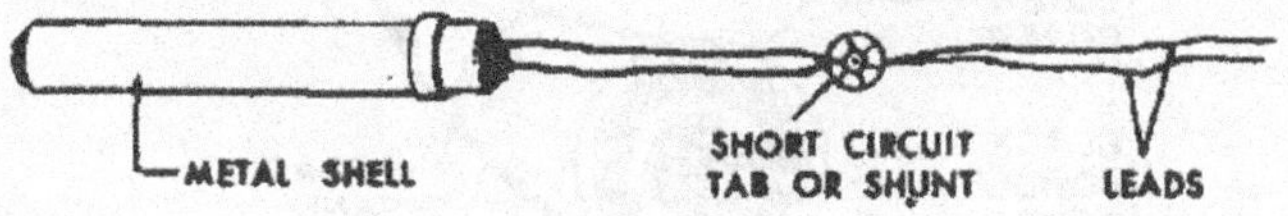

q. Nonelectric Blasting Caps. Two types are available, the No. 8 and the special M7, which resembles the No. 8 in appearance except for the expanded open end.

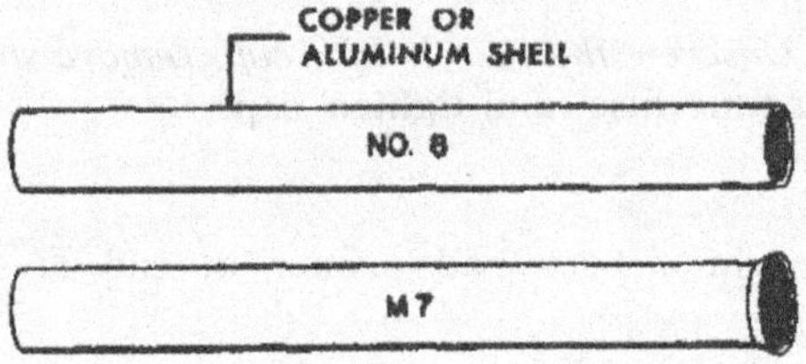

. *Priming Adapter.* This is a plastic device with a threaded end for securing electric and nonelectric primers in the threaded cap wells of military explosives. A groove for easy insertion of the electric lead wires extends the full length of the adapter.

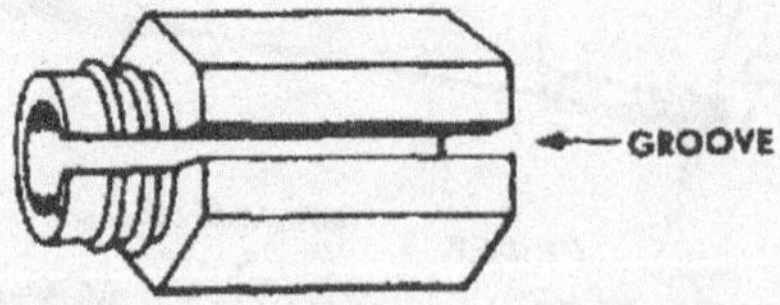

. *M10 Universal Destructor.* The destructor is used to convert loaded projectiles, missiles, and bombs into improvised charges. The destructor has booster caps containing tetryl pellets. All standard firing devices with the standard base coupler screw into the top.

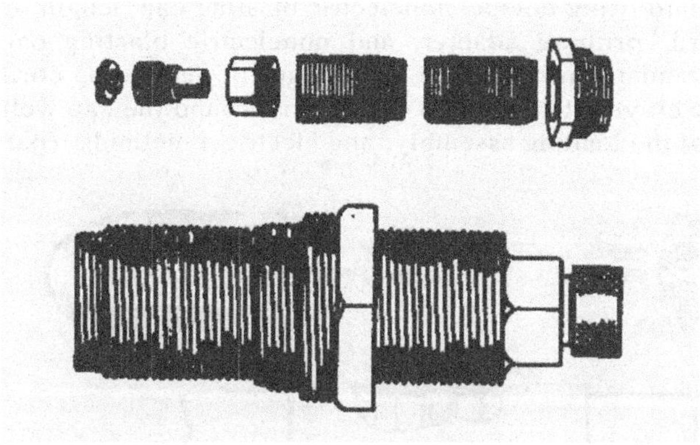

t. ***Antitank Mine Activator.*** This is a detonator designed for boobytrapping antitank mines. The top is threaded to receive all standard firing devices, and the base to screw in antitank mine activator wells.

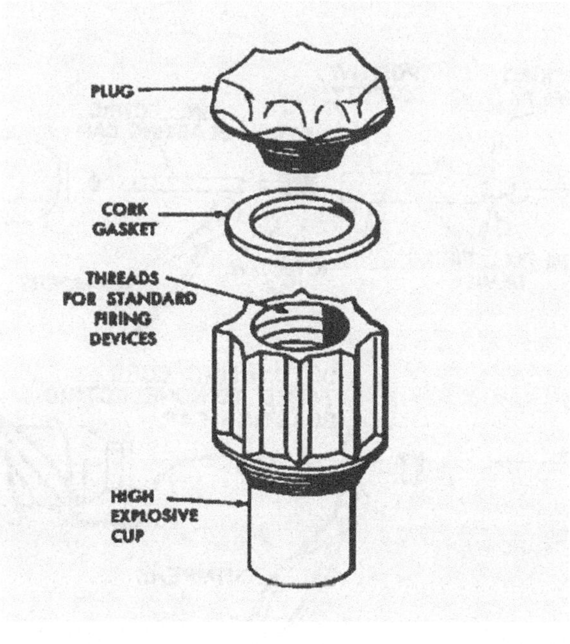

28. Bangalore Torpedo

The bangalore torpedo is a group of 10 loading assemblies (steel tubes filled with high explosive) with nose sleeve and connecting sleeves. The loading assemblies may be used singly, in series, or in bundles. They are primed in four ways: by a standard firing device; a standard firing device, nonelectric blasting cap, length of detonating cord, priming adapter, and nonelectric blasting cap (para 29); a standard firing device, and length of detonating cord attached by the clove hitch and two extra turns around the cap well at either end of the loading assembly; and electrical methods (para 29).

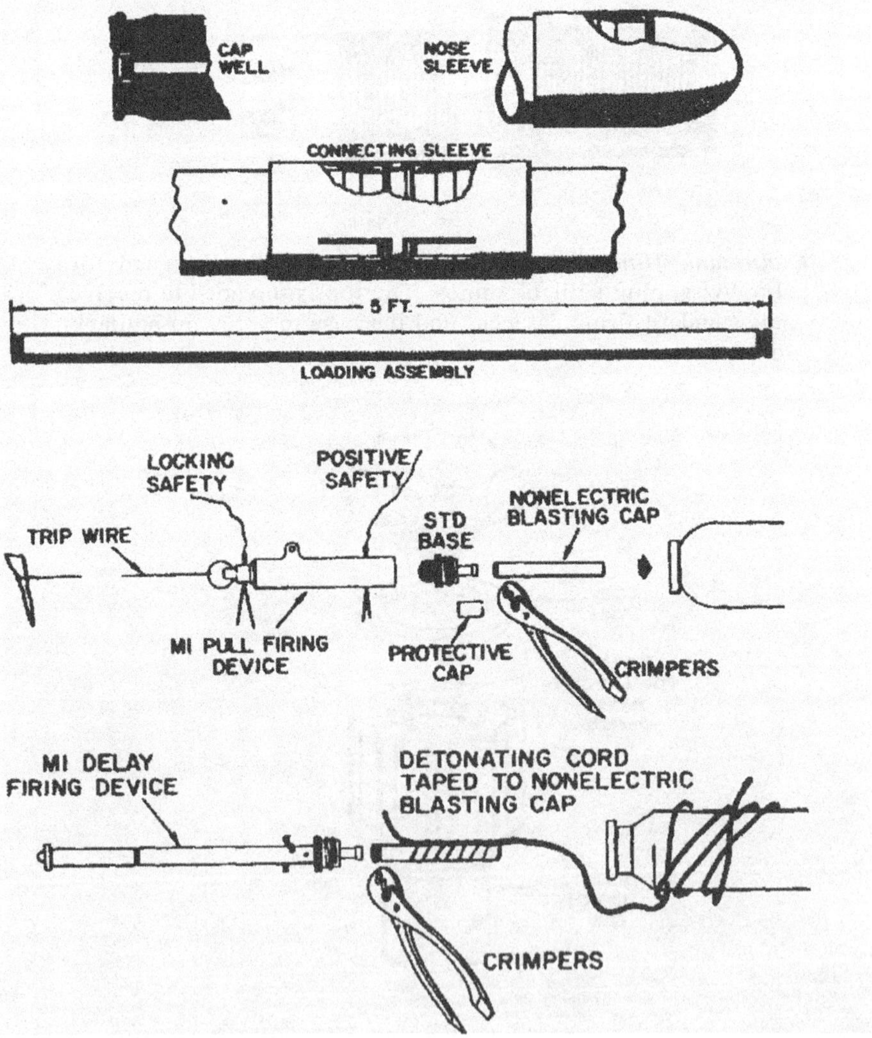

29. M2A3 Shaped Charge

This charge consists of a conical.top, conical liner, integral standoff, threaded cap well, and pounds of explosive. It may be primed in three ways: by a standard firing device; a standard firing device, nonelectric blasting cap, length of detonating cord, priming adapter, and nonelectric blasting cap; and a priming adapter and electric blasting cap connected to power source.

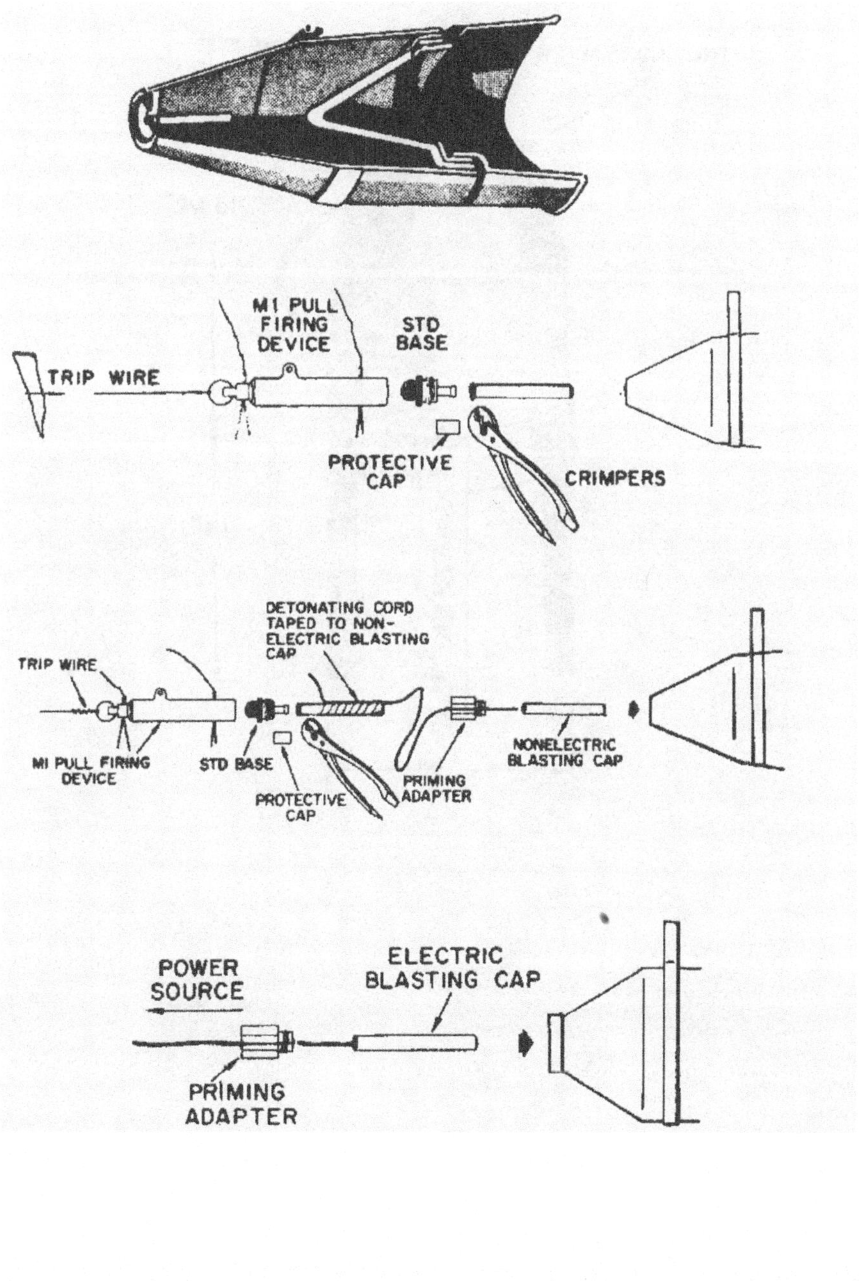

30. M3 Shaped Charge

The M3 shaped charge is a metal container with a conical top, conical liner, threaded cap well,' 30 pounds of explosive, and a metal tripod standoff. It may be primed in the same manner as the M2A3 shaped charge above.

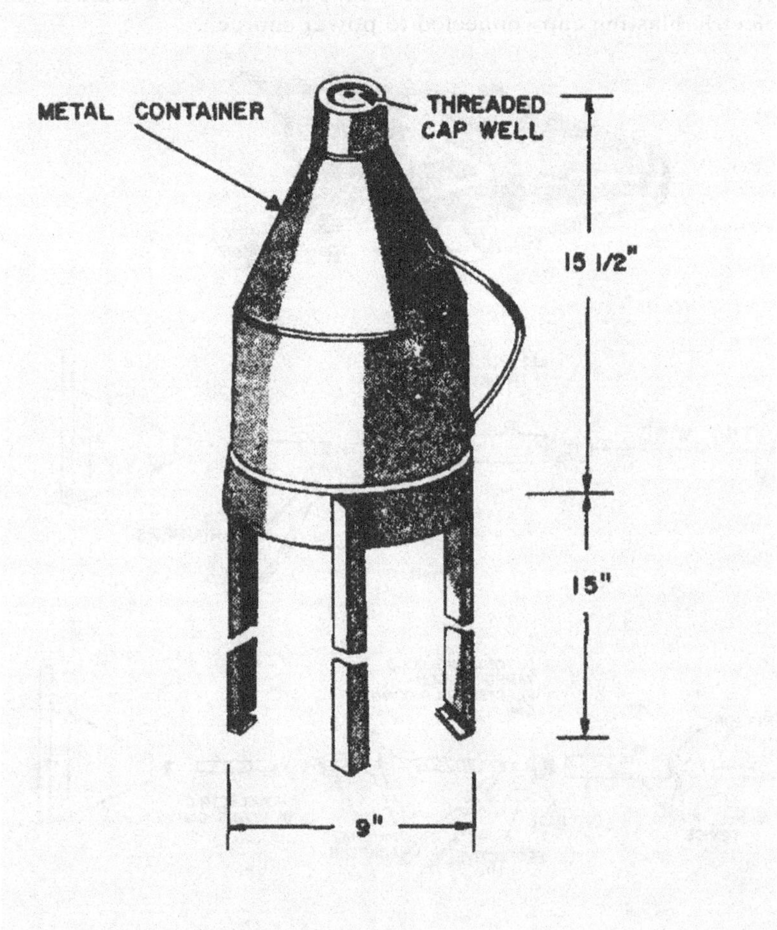

Section III. MISSILES

31. Introduction

Hand grenades, bombs, and mortar and artillery ammunition have wide application as improvised explosive charges. The only portion of these useful in boobytrapping, however, are the container and its explosive filler. The fuze is replaced by a standard firing device and an M10 universal destructor—an adapter designed especially for this purpose. The number and type of missiles useful in boobytrapping, however, are not limited to the examples given below.

The M26 hand grenade, an improved model, consists of a thin metal body lined with a wire-wound fragmentation coil, fuze, and composition B explosive charge. It has a variety of applications to boobytrapping. The fuze is removed and a standard firing device is screwed directly into the fuze well or remotely connected by a length of detonating cord, priming adapter, and a nonelectric blasting cap.

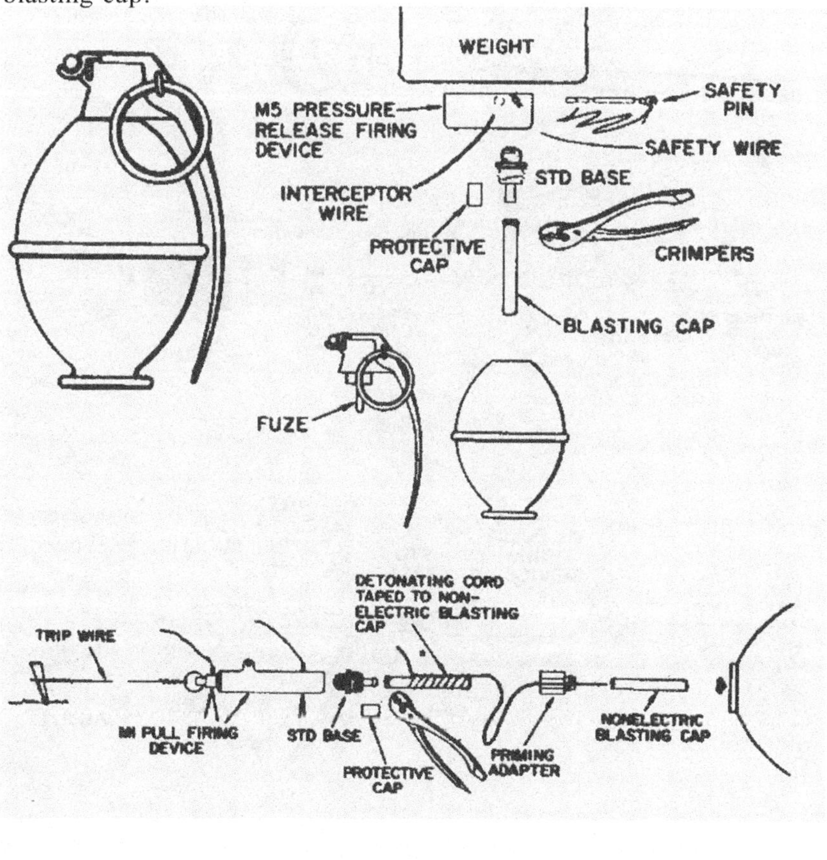

33. 81MM Mortar Sholl

This is converted by replacing the fuze with a standard firing device and a properly assembled 'destructor or by a firing device, length of detonating cord, priming adapter, nonelectric blasting cap, and a properly assembled destructor. If a destructor is not available the detonating cord and nonelectric blasting cap are packed firmly in the fuze well with C4 explosive.

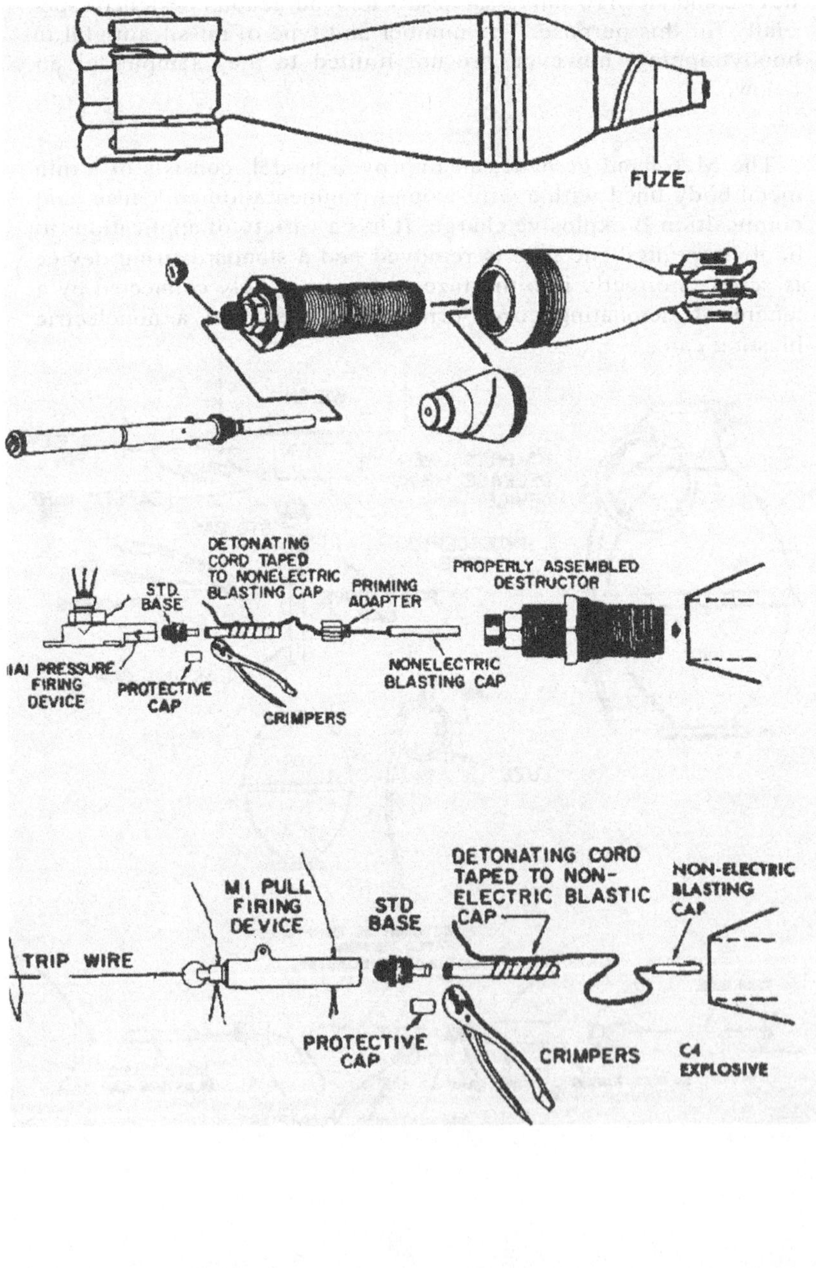

34. High Expletive Shell

The high explosive shell is readily adapted to boobytrapping. The fuze is removed and replaced by'a standard firing device and a properly-assembled destructor or a standard firing device, length of detonating cord, priming adapter, nonelectric blasting cap, and a properly-assembled destructor. If a destructor is not available, the detonating cord and nonelectric blasting cap are packed firmly in the fuze well with C4 explosive.

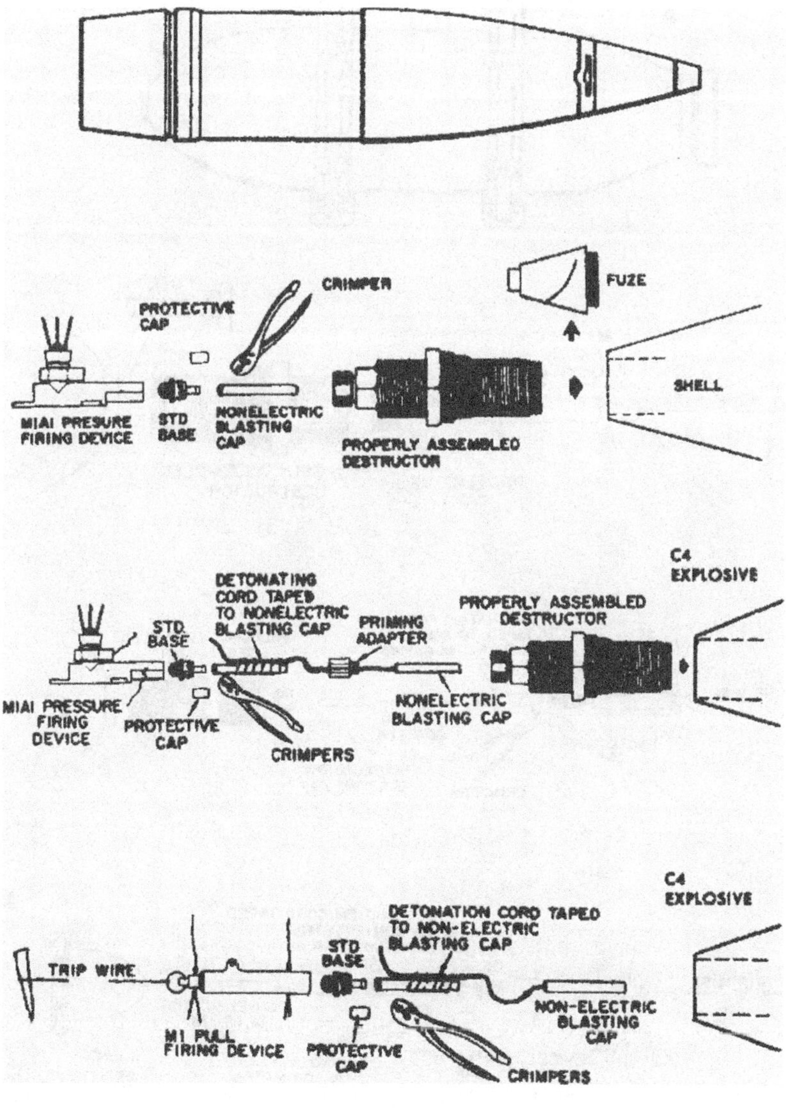

35. Bombs

These are adapted to boobytrapping in the same manner as high explosive and mortar shells. They are primed by replacing the fuze with a standard firing device and a properly-assembled destructor, or with a standard firing device, length of detonating cord, priming adapter, nonelectric blasting cap, and a properly-assembled destructor. If a destructor is not available, the detonating cord and blasting cap are packed firmly in the fuze well with C4 explosive.

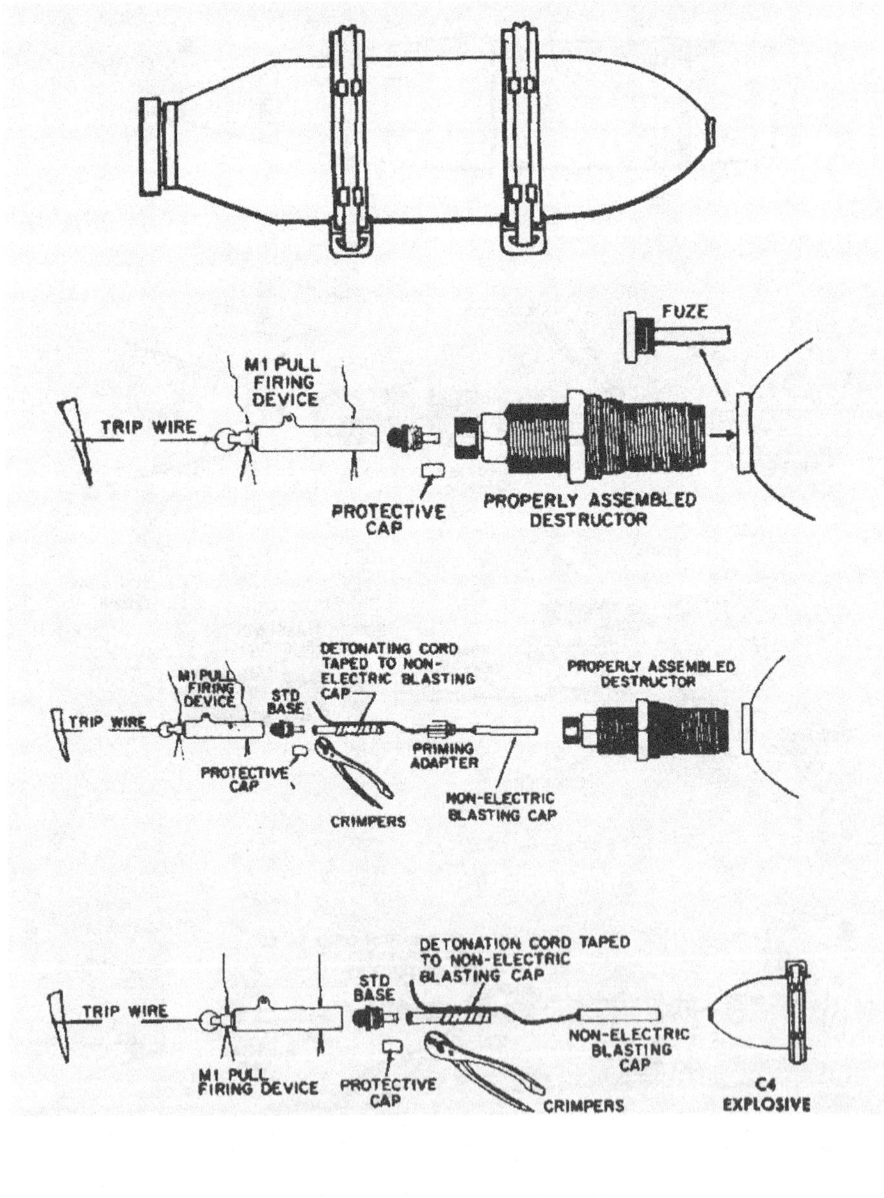

36. Antitank Mines

A land mine may be used as the main charge in a boobytrap by removing the fuze and attaching a standard pull or pressure-release firing device in an auxiliary fuze well.

(1) Remove locking safety cotter pin in Ml pull firing device and replace with length of thin wire. Bend wire slightly to prevent dropping out.
(2) Remove positive safety cotter pin and replace with length of thin wire. Bend wire slightly to prevent dropping out.
(3) Remove plastic protective cap from standard base.
(4) Assemble firing device, activator, and mine.

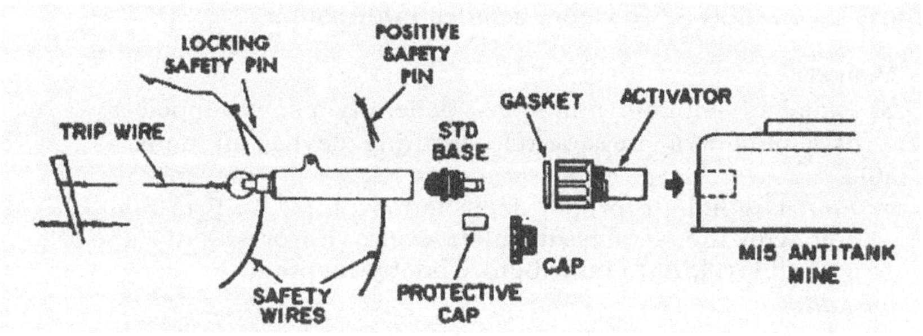

6. *Pressure-Release.*

(1) Insert length of heavy wire in interceptor hole in M5 pressure-release firing device. Bend wire slightly to prevent dropping out.
(2) Withdraw safety pin and replace with length of thin wire. Bend wire slightly to prevent dropping out.
(3) Remove plastic protective cap from standard base.
(4) Assemble firing device, activator, and mine.

Note. The firing device must be set on a firm base. A piece of masonite is issued with the M5 for this purpose.

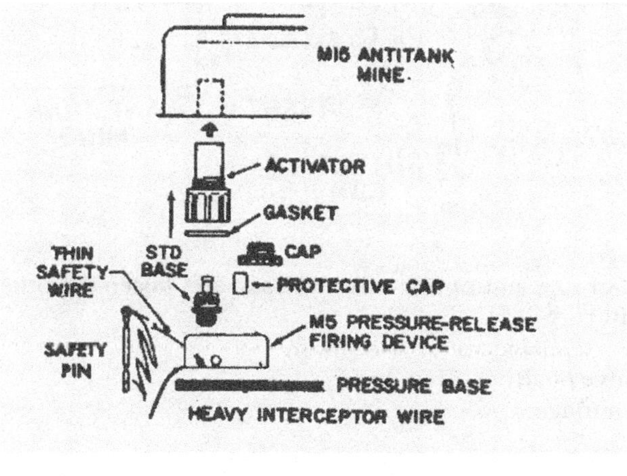

CHAPTER 4

CONSTRUCTION TECHNIQUES

Section I. Boobytrapping Minos in Minefields

37. Tactical Purpose

Antitank mines laid in mine fields are boobytrapped (or activated) primarily to make breaching and clearing as dangerous, difficult, and time consuming as possible in order to confuse, demoralize, and delay the enemy. Most standard U.S. antitank mines and many foreign antitank mines have auxiliary fuze wells for this purpose. See FM20-32 for more detailed information.

38. Mothods

U.S. standard antitank mines are generally boobytrapped by means of a pull or a pressure-release firing device, or both, if desirable.

a. Pull. Dig hole to proper depth to bury mine on firm foundation with top of pressure plate even with or slightly above ground level. Arm mine before boobytrapping.

(1) *Installing.*

(a) Remove locking safety cotter pin and replace with length of thin wire. Bend wire slightly to prevent dropping out.

(b) Remove positive safety cotter pin and replace with length of thin wire. Bend wire slightly to prevent dropping out.

(c) Remove protective cap from standard base and assemble firing device, activator, and mine.

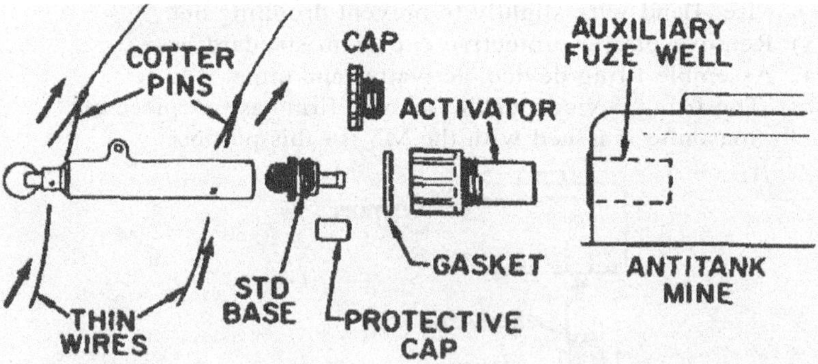

(2) *Arming.*

(a) Anchor one end of trip wire to stake and fasten the other to pull ring.

(b) *Remove locking safety wire first.*

(c) Remove positive safety but.

(d) Camouflage.

(3) *Disarming.*

(a) Uncover mine carefully.
(b) Locate boobytrap assembly.
(c) Replace positive safety *first,* then locking safety.
(d) Cut trip wire.
(e) Turn arming dial of mine to *safe* and remove arming plug.
(f) Remove fuse and replace safety clip.
(g) Replace arming plug.
(h) Recover mine and firing device.

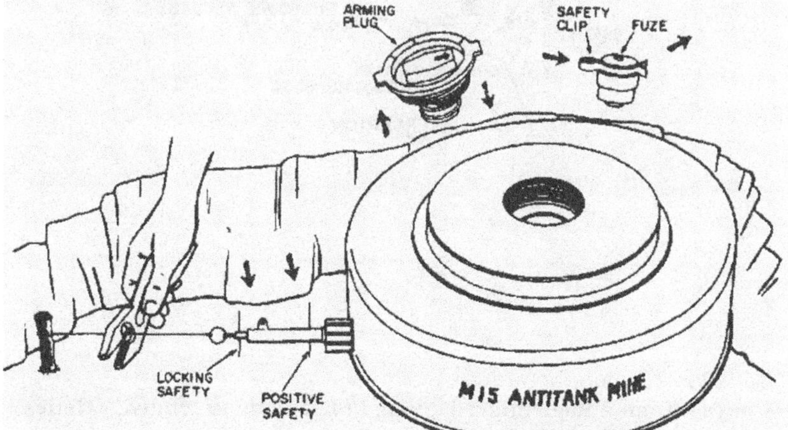

b. Pressure-Release. Dig hole to proper depth to bury mine on firm foundation, with top of pressure plate even with or slightly above ground level. .

(1) *Installing.*

(a) Insert length of heavy wire in interceptor hole. Bend wire slightly to prevent dropping out.
(b) Remove safety pin. Apply pressure on release plate until pin comes out easily.
(c) Insert length of light wire in safety pin hole and bend slightly to prevent dropping out.
(d) Remove protective cap from standard base and assemble firing device, activator, and mine.
(e) Place mine and firing assembly in hole, using pressure board to insure a solid foundation for firing device.

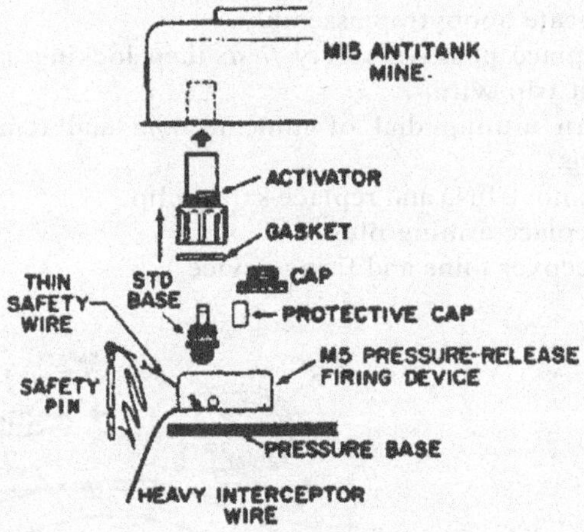

(2) *Arming.*
(a) Camouflage mine, leaving hole at side to remove safeties.
(b) Carefully remove thin safety wire *first,* then the interceptor wire.
(c) Complete camouflage.

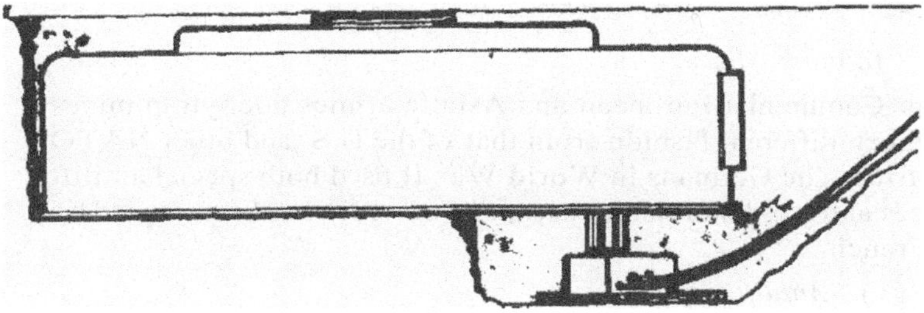

(3) *Disarming.*
(a) Uncover mine carefully.
(b) Locate boobytrap assembly.
(c) Insert length of heavy wire in interceptor hole.
(d) Turn dial on pressure plate to "S" (safe) and replace safety fork.
(e) Recover mine and firing device assembly.
(f) Remove pressure plate, unscrew detonator, and replace shipping plug.
(g) Reassemble mine.

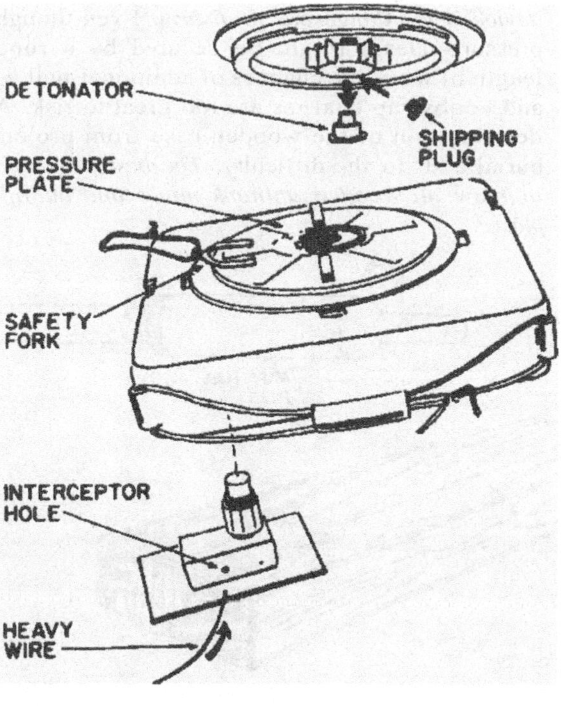

39. Boobytrapped Foreign Mines

a. Antitank Mines.

The Communist European and Asiatic armies boobytrap mines in a much different fashion from that of the U.S. and other NATO countries. The Germans in World War II used both special antilift devices and antidisturbance fuzes, one of which has been copied by the French.

(1) *Antilift devices,*

(a) Russia

1. The Russians, Communist Chinese, and North Koreans boobytrapped wooden antitank mines by laying two of them, one on top of the other, in the same hole. The mines were connected by an MUV pull fuze and a pull wire, so that the bottom mine would detonate when the top mine was lifted.

2. The Russians in World War II also had a more sophisticated method—a special wooden antilift device, placed under the mine. This, however was readily located by probing. It consisted of an outer case, a charge, an MUV pull fuze, a pressure release lid supported on two coil springs, and a fuze access hole. Lifting the mine initiated the antilift. *This device is too dangerous to disarm.* Even though the pressure-release might be secured by a rope or length of wire, the chances of additional pull wires and boobytrap charges are too great to risk. Also deterioration of the wooden case from prolonged burial adds to the difficulty. *The best procedure is to blow all wooden antitank mines and antilifts in place.*

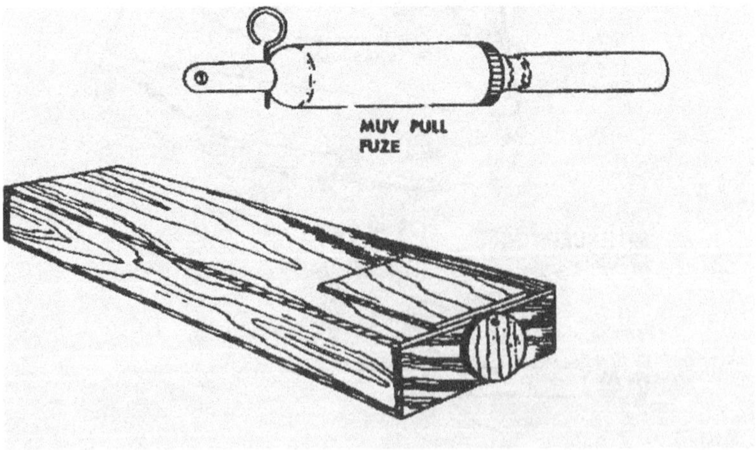

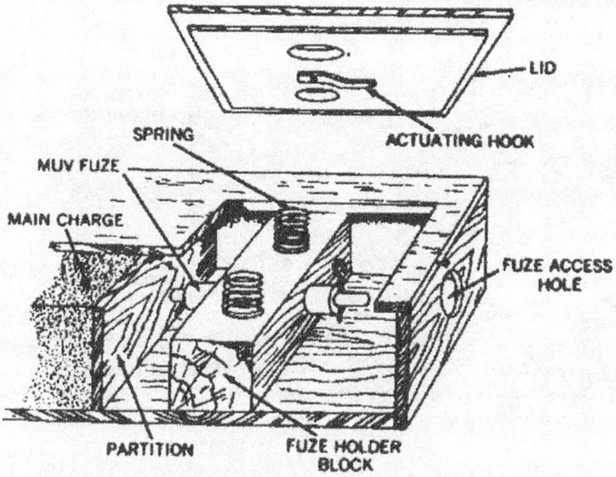

(b) *Czechoslovakia.* This satellite country has a wooden antitank mine (PT-Mi-D) that may prove extremely hazardous to breaching and clearing parties. Having an RO-1, pull fuze in each end, it is easily boobytrapped by means of wire anchored to a stake underneath the mine and extended through a hole in the bottom of the case to the fuze pull pin.

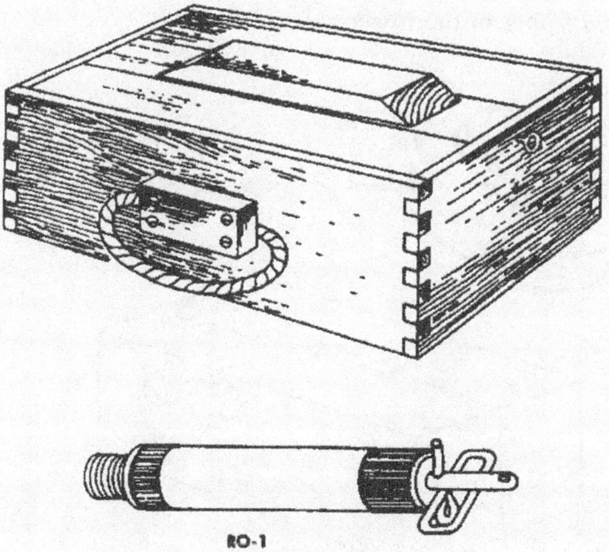

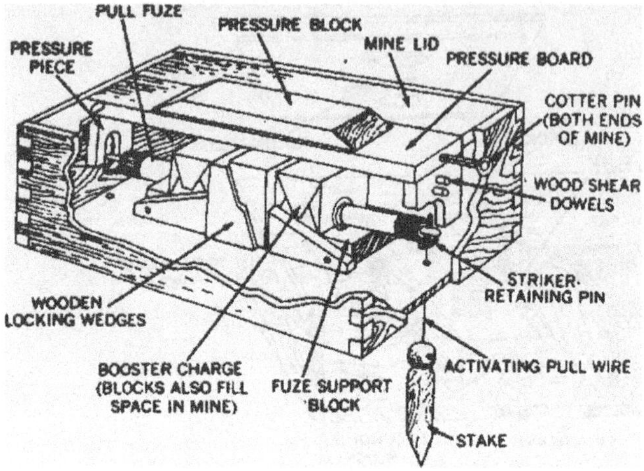

(c) *World War II Germany.* The German armies had several pressure-release devices for boobytrapping antitank mines. In a future war in Europe, these or facsimiles may appear on any battlefield.

1 *Nipolite all explosive antilift.* This consisted of two oblong blocks of moulded explosive joined together with brass bolts and recessed to contain the metal striker assembly. It may be disarmed by inserting a safety in the lower safety pin hole.

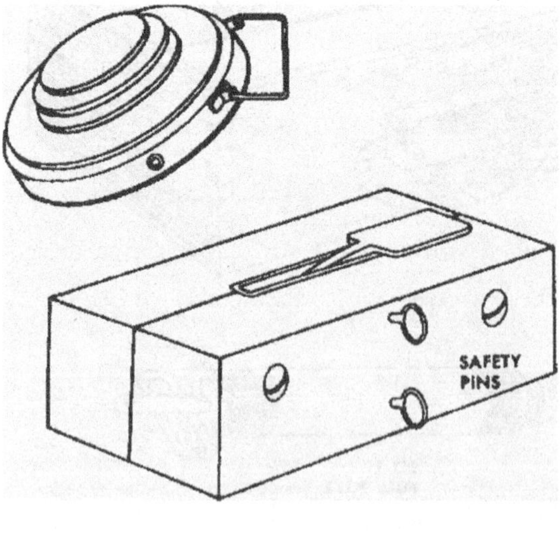

t *EZ. SM2 (EZ U).* This device consists of an explosive charge, a pressure-release firing mechanism, a safety bar and a metal case. When the safety bar is removed, the device arms itself by means of clockwork inside the case. *This device cannot be disarmed.*

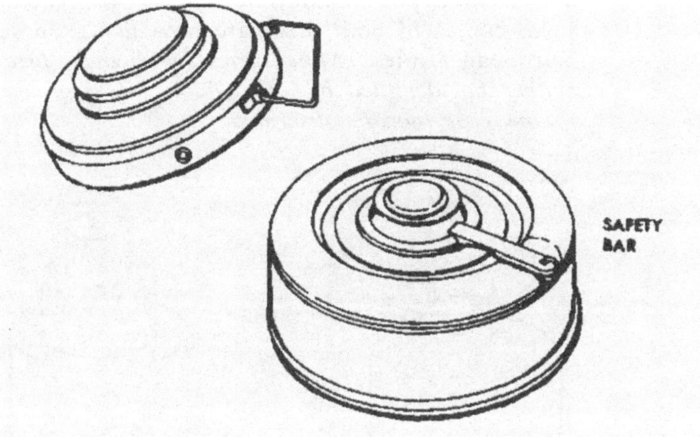

S *SFS.* This antilift consists of an explosive charge, pressure-release striker assembly, safety bar, and chemical arming equipment. A turn of the safety bar crushes the glass vial, releasing the,chemical' to dissolve the safely pellet. *This device cannot be disarmed.*

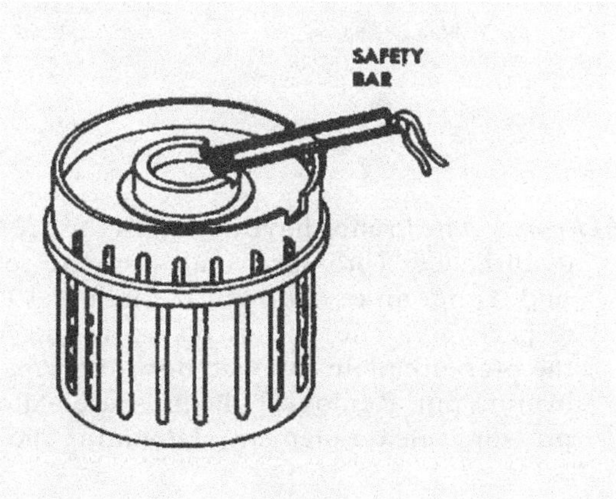

(2) ***T. Mi. Z 4S and T. Mi Z 44 antidisturbance fuzes.***

(a) *Germany.* In addition to several antilift devices, the Germans developed two antidisturbance fuzes initiated by pressure or pressure-release for activating Teller mines 42 and 43. To arm, the fuze is placed in the fuze well and the pressure plate screwed down on top of the fuze, shearing the arming pin. Removal of the pressure plate initiates the pressure-release mechanism and detonates the mine. Although the T. Mi. Z 44 was an experimental model that never reached the field, copies of both fuzes are now in use in several European armies. *Mines armed with these fuzes can neither be identified by size, shape, marking, or color of the case, nor be disarmed.*

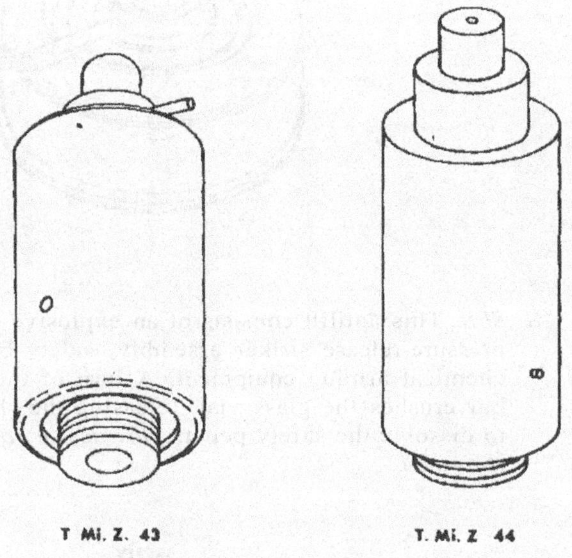

(b) *France.* The French have a copy of the T. Mi. Z 43 antidisturbance (pressure and pressure-release) fuze, and Teller mine 43, named models 1952 and 1948 respectively. The fuze is placed in the fuze well and the pressure plate screwed down on top, shearing the arming pin. Removing the pressure plate actuates the pressure-release element, detonating the mine.

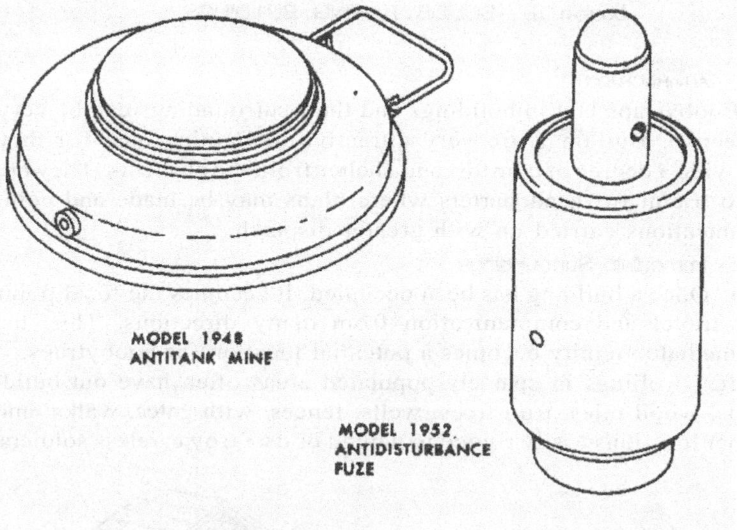

b. Antipersonnel Mines.

Antipersonnel mines are laid in antitank minefields to halt and delay enemy troops and make breaching and clearing as difficult, dangerous, and time consuming as possible. Enemy mine layers may increase this harrassment substantially by laying small blast type antipersonnel mines near the anchors and along the trip wires, which, according to procedure, must be traced from pull ring to anchor before cutting. These are extremely hazardous to breaching and clearing specialists who may detonate them unawares by the pressure of a hand, knee, or elbow on the pressure plate.

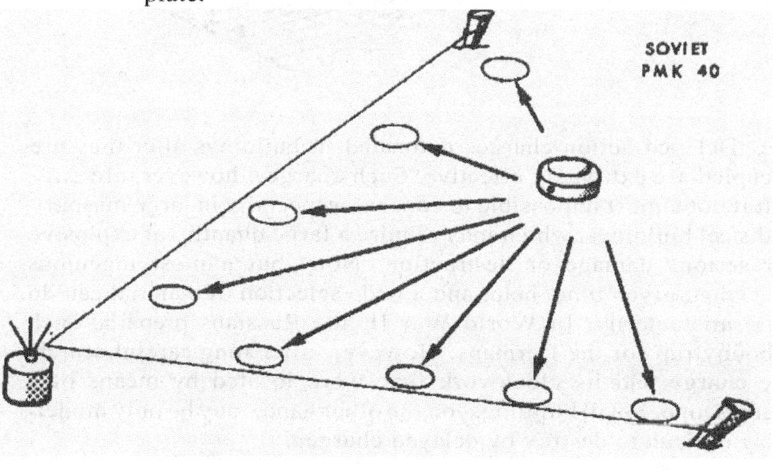

Sodic-n II. BCMDBYTfAPfING BUILDINGS

40. AchnHflfCHJOS

Boobytraps laid in buildings and their surroundings can be very effective. Buildings are very attractive to fighting men for they provide a degree of comfort and shelter from the elements. They are also useful for headquarters where plans may be made and communications carried on with greater dispatch.

41. Immodiato Surroundings

a. Once a building has been occupied, it becomes the focal point for travel and communication from many directions. Thus the immediate vicinity becomes a potential location for boobytraps.

ft. Dwellings in sparsely populated areas often have out buildings, wood piles, fruit trees, wells, fences with gates, walks, and other locations easily rigged to wound or d< • troy careless soldiers.

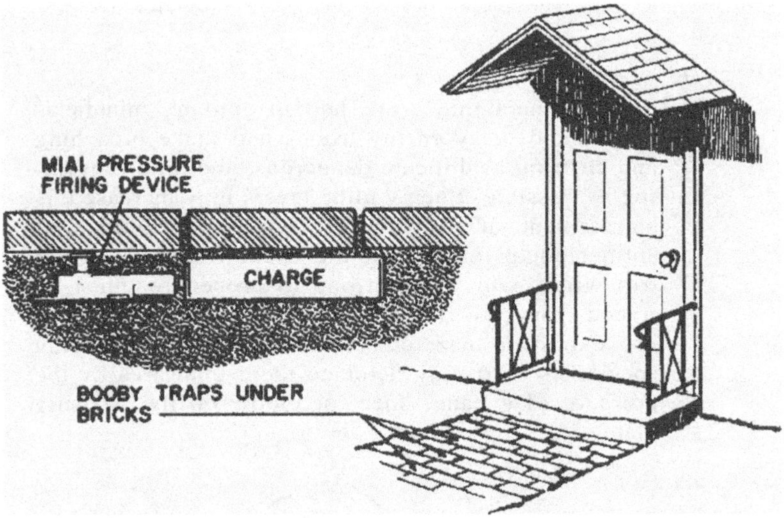

c. Delayed action charges detonated in buildings after they are occupied are extremely effective. Such charges, however, are difficult if not almost impossible to conceal, especially in large masonry and steel buildings, which may require a large quantity of explosive for serious damage or destruction. None but a most ingenious specialist, given time, help, and a wide selection of material can do this satisfactorily. In World War II, the Russians prepared such a boobytrap for the Germans. However, after long careful search, the charge and its clockwork fuze were located by means of a stethoscope. Small buildings, on the other hand, may be only moderately difficult to destroy by delayed charges.

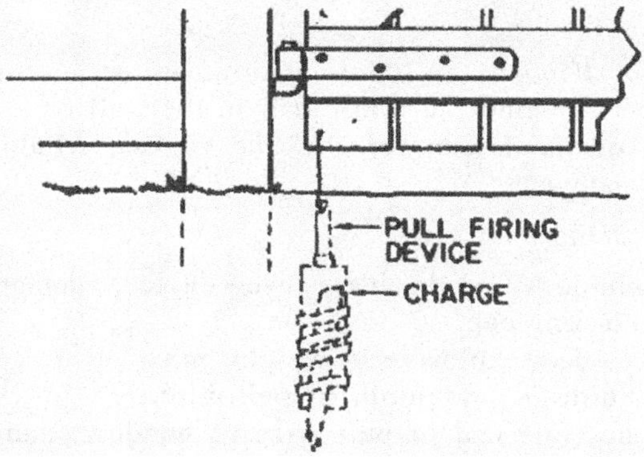

42. entrances

Curiosity prompts a soldier to investigate hurriedly an interesting building in his path. Women, loot, or mere inquisitiveness **may** be the motive. His rush to be the first inside makes all entrances excellent spots for boobytraps. For the foolish, a rigging connected to the front door, side door, or back doors may be sufficient But for the experienced soldier, who may carefully seek entry to the basement first and then try to clear the building story by story, careful and ingenious effort may be required.

a. Basement Windows. Here boobytraps must be concealed to prevent detection by the enemy's breaking the pane or kicking out a door panel. Basement windows should be boobytrapped at the top or in the floor underneath.

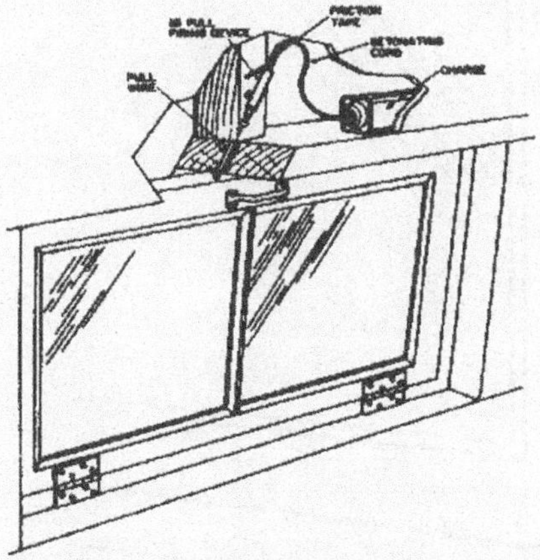

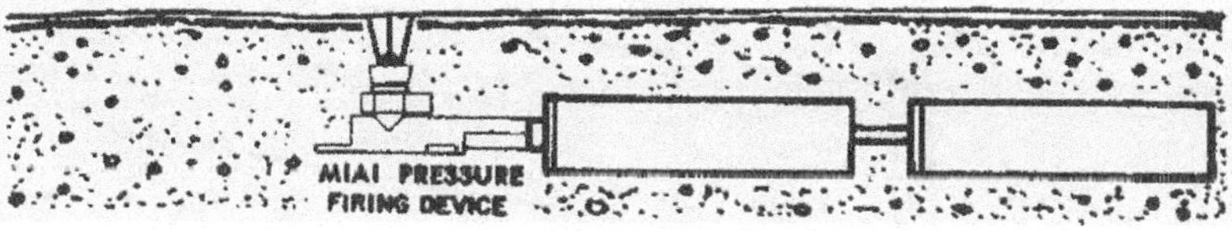

b. Upper Floor Windows. Window charges are easier concealed in the weight box behind the jamb than in the wall or under the floor. Experienced hands can remove and replace window trim without obvious damage.

(1) *Nonelectric firing.*

(a) Assemble M3 pull-release firing device, standard base, and blasting cap.
(b) Place sheet explosive in weight box.
(c) Bore hole in side jamb for pull wire.
(d) Anchor one end of pull wire to window, and thread through hole in side jamb.
(e) Attach free end of pull wire to ratchet on firing device.
(f) Arm firing device.
(g) Conceal boobytrap.

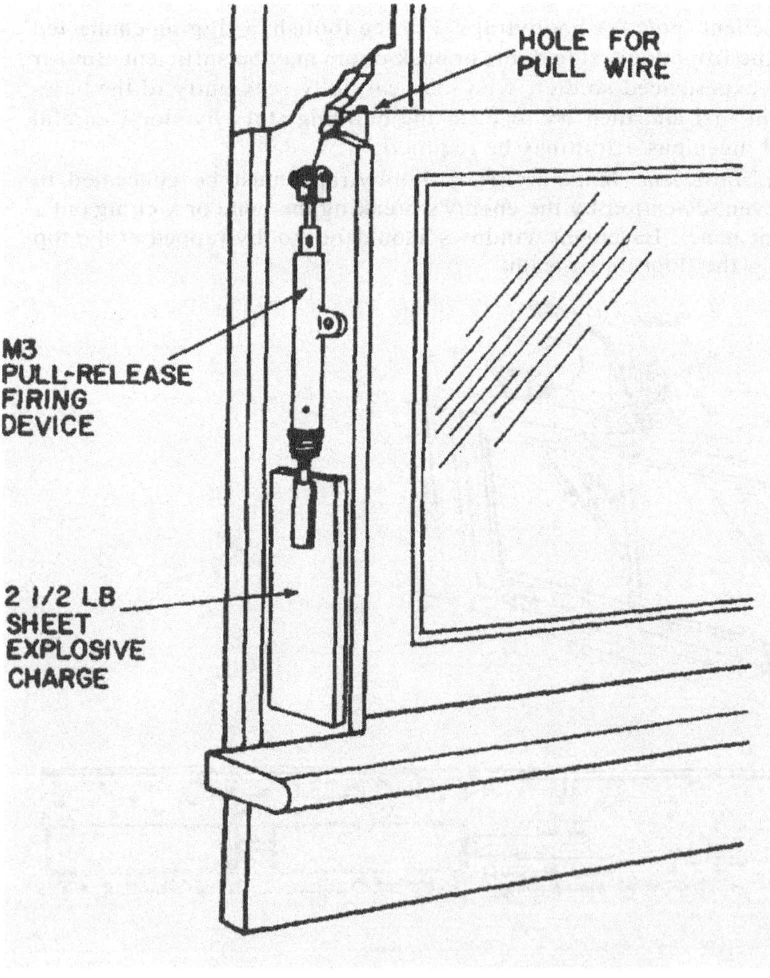

(2) ***Electric firing.***

(a) Fasten two metal brackets to side of weight box close enough to wedge two flashlight batteries between.
(b> Place sheet explosive charge in weight box.
(c) Insertrelectric blasting cap in charge.
(d) Cut one leg wire and attach to lower bracket.
(e) Cut other leg wire to proper length to twist an uninsulated loop on end and fasten to hang in .place just above top of window weight.
(f) On a length of leg wire twist on uninsulated loop around the leg wire hanging above the weight. Thread other end through other uninsulated loop and fasten to top clamp. Tape wire to window weight.
(g) Test circuit with galovonmeter first, then insert batteries between brackets.
(h) Conceal boobytrap.

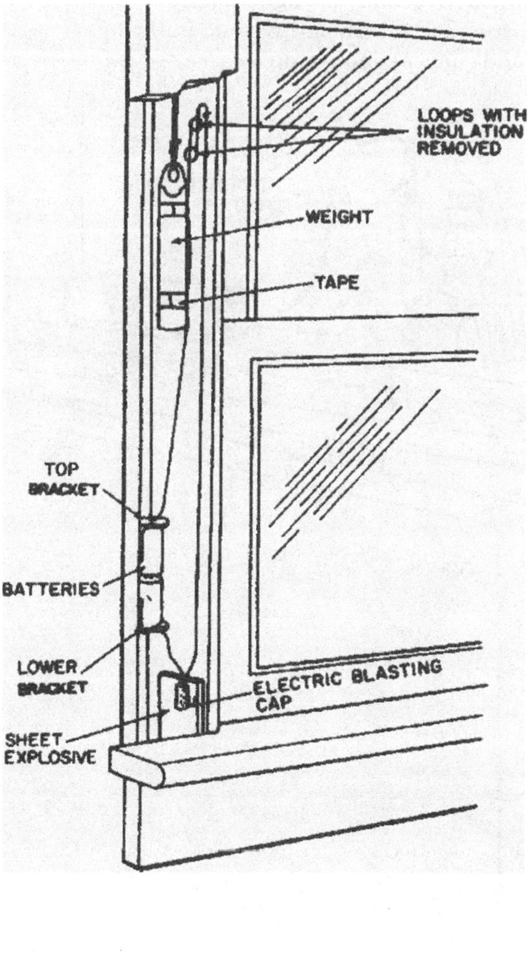

e. Doors. Improved detection methods have made the use of boobytraps on doors, with charges, firing devices, and wires exposed, a waste of time and material, except for purposes of deception. The best location is the head or side jamb, not the sill, which is often recommended. The sill is exposed, so that one experienced clearing unit may easily locate the rigging while in the jamb, it is concealed by the doorstop.

(1) *Head jamb rigging.*

(a) Assemble Ml pull firing device, standard base, and nonelectric blasting cap.

(b) Assemble length of detonating cord, priming adapter, nonelectric blasting cap and explosive block.

(c) Attach firing device firmly to stud and tape free end of length of detonating cord to nonelectric blasting cap.

(d) Drill hole at proper place in header and head jamb.

(e) Anchor one end of pull wire at proper place on door and thread free end through holes.

(f) Close door and attach pull wire to pull ring.

(g) Arm and conceal boobytrap.

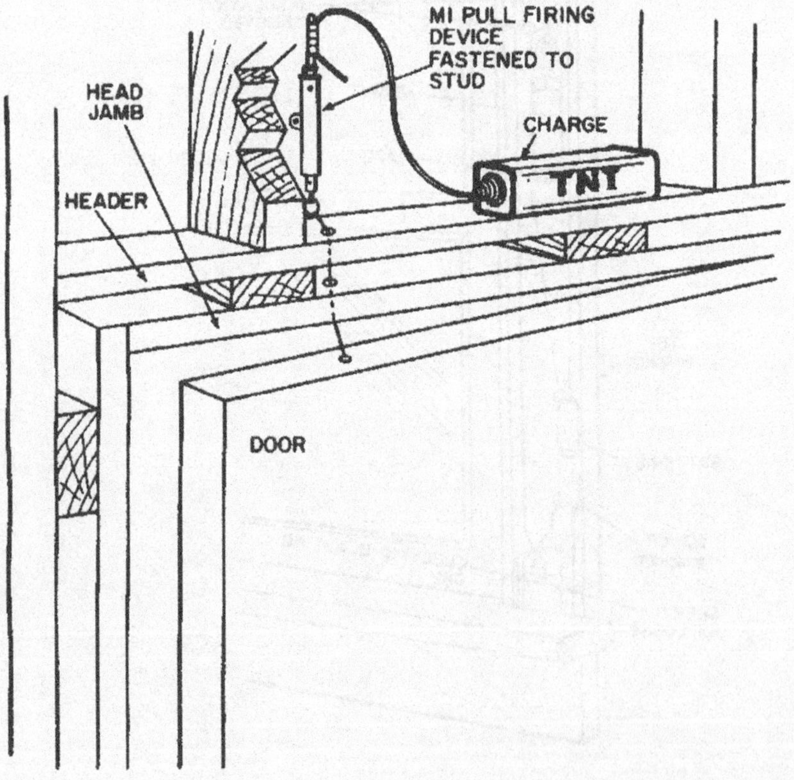

(2) ***Side jamb rigging.***

(a) Attach metal brackets to side jamb close enough to wedge two flashlight batteries between.

(b) Insert sheet explosive charge snugly between stud and jamb.

(c) Place electric blasting cap in charge, and fasten one leg wire to top bracket

(d) Bore pull wire hole at proper spot inside jamb.

(e) Cut other leg wire long enough to twist on an insulated loop on one end and fit over pull wire hole. Loop should be about ^ inch in diameter.

(f) Twist on uninsulated loop on one end of leg wire and secure to lower bracket so that loop fits over pull wire hole. Fasten wire to jamb.

(g) Anchor one end of insulated pull wire at proper spot on door, and thread free end through pull wire hole and loop fastened to jamb.

(h) Close door. Fasten free end of pull wire to other loop to hold it snugly against stud.

(i) Check circuit with galvonometer first, then

(j) Install batteries between brackets.

(k) Conceal boobytrap.

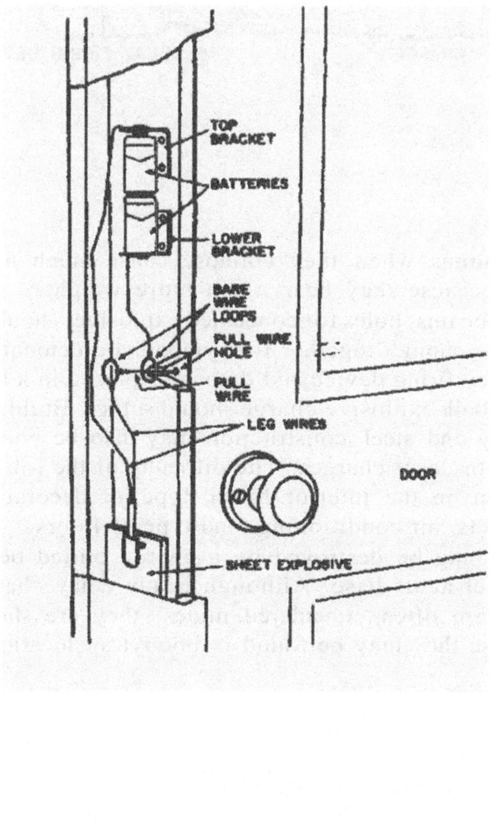

43. Structural Framework

a. In a building charges should be placed where detonation will seriously impair its structural strength, such as walls, chimneys, beams, and columns. Charges and firing devices must be carefully concealed to avoid detection.

6. In boobytrapping load-bearing walls, several charges should be laid to detonate simultaneously near the base. Chimneys and fireplaces are difficult to boobytrap for charges placed there are readily detected. These should detonate from intense heat.

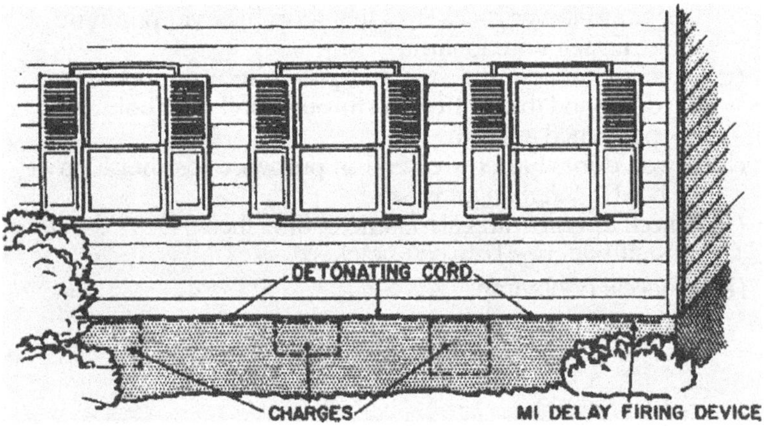

e. Beams and columns when they collapse cause much more damage than walls because they bear much more weight.

(1) In wooden beams, holes for concealed explosives should be bored close enough together for sympathetic detonation. An Ml delay firing device and detonator placed in a hole within the bulk explosive charge should suffice. Buildings of masonry and steel construction may also be boobytrapped with delay charges. The difficulty of the job depends often on the interior finish, type of decoration, heating ducts, air conditioning, and type of floors.

(2) A column may be destroyed by a charge buried below ground level at its base. Although heavy delay charges like these are often considered mines, they are shown here because they may be found in boobytrap locations.

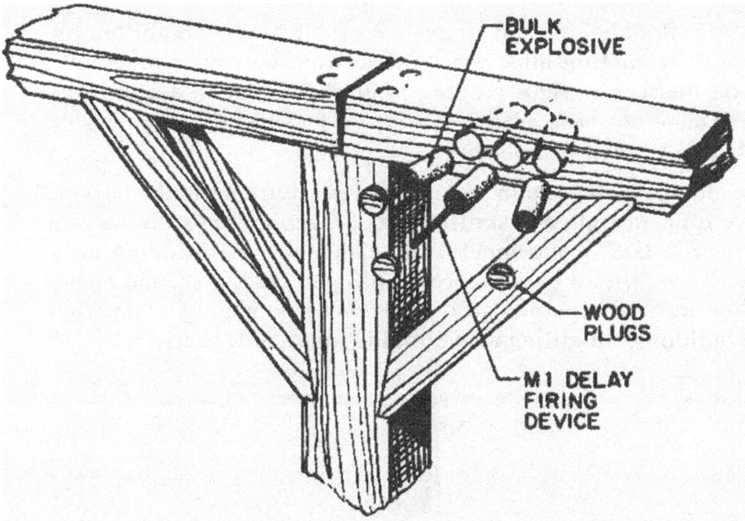
BULK EXPLOSIVE
WOOD PLUGS
M1 DELAY FIRING DEVICE

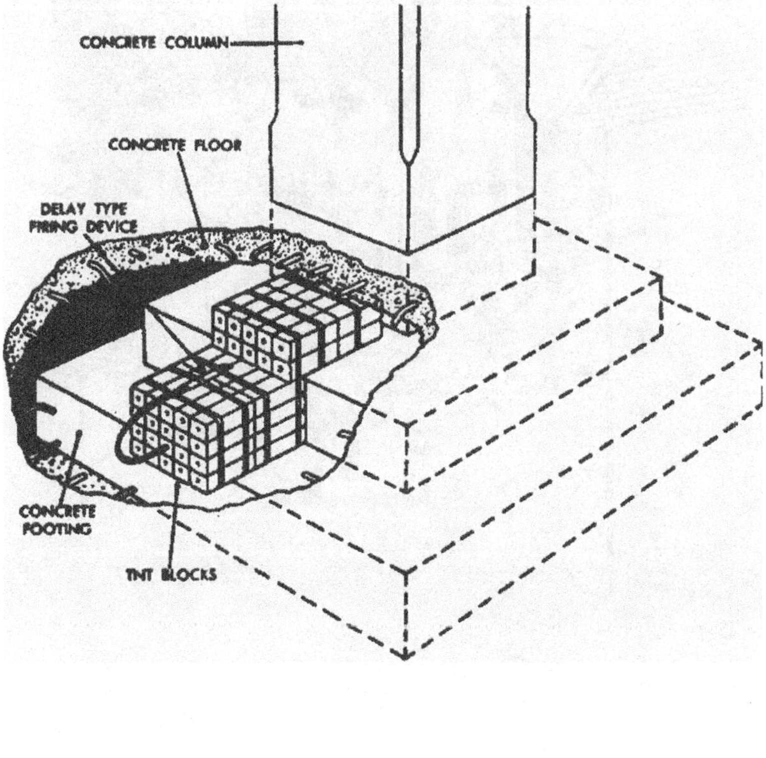
CONCRETE COLUMN
CONCRETE FLOOR
DELAY TYPE FIRING DEVICE
CONCRETE FOOTING
TNT BLOCKS

d. Loose floor boards sometimes are excellent objects for booby-trapping. The rigging must escape detection, however; otherwise, it will be ineffective. This rigging might be harder to detect if the support underneath is chiseled out to let the floorboard sink about 1/4 inch when tramped on.

e. A double delay chain detonating boobytrap should be very effective if timed right and skillfully laid. *First,* is the explosive of a minor charge laid in an upper story damaging the building only slightly. *Then,* after a curious crowd has gathered, a second heavy charge or series of charges go off, seriously damaging or destroying the building and killing or wounding many onlookers.

44. Interior Furnishings

Vacated buildings provide much opportunity for boobytrapping. Hurriedly departing occupants usually leave behind such odds and ends as desks, filing cases, cooking utensils, table items, rugs, lamps, and furniture. Electric light and power fixtures are also exploitable.

a. *Desk.* Because of its construction a desk is easily boobytrapped. If carefully placed the rigging may be nondetectable and if properly constructed, cannot be neutralized. Electric firing systems are the most suitable for this purpose. Sheet explosive is much better than other types, because its adhesive surface holds it firmly in place. Check the circuit with a galvonometer *before* installing the batteries.

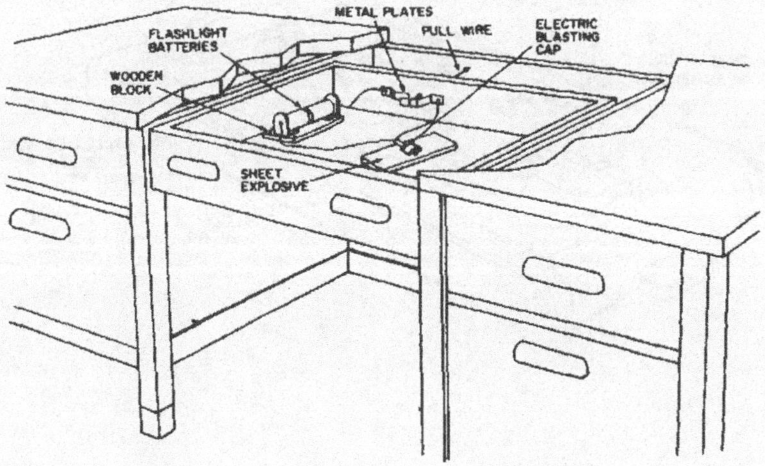

6. *Office Equipment.* Many items used in offices have boobytrap potential.

(1) *Telephone list finder.*

(a) Remove contents from finder.
(b) Assemble sheet explosive, shrapnel, and blasting cap.
(c) Remove insulation from ends of wires and twist to form loop switch.
(d) Place boobytrap in finder so that the raising of the lid draws the loops together.
(e) Insulate inside of case from contact with loops with friction tape.

(f) Check circuit with galvanometer *first,* then install batteries.

Note. Batteries may be connected to legwires by wrapping them tightly in place with friction tape.

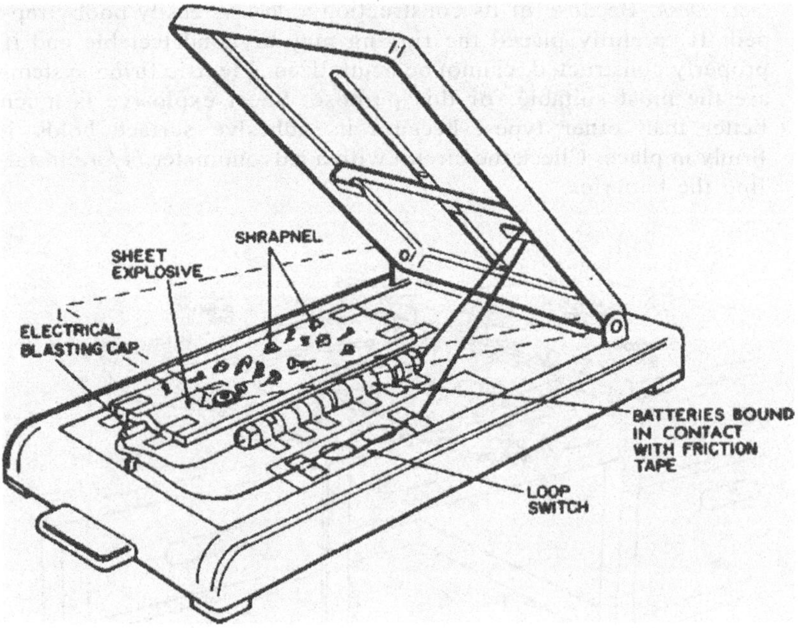

(2) *Card File.* A wooden card file can be boobytrapped effectively by the use of a mousetrap rigged as a trigger, a standard base with blasting cap attached, a support block fastened inside to hold the firing assembly at the proper level for operation, and a trigger block to hold the trigger in armed position.

(a) Rig wire trigger of mousetrap with screw and metal strip.

(b) Locate support block on strips at proper level to fix trigger in trigger block.

(c) Bore hole in support block at proper place to admit standard base and blasting cap so that sheet metal screw will strike percussion cap.

(d) Insert explosive, then support block with mousetrap, standard base, and blasting cap in position.

(e) Raise trigger and close lid so that trigger is fixed in firing position.

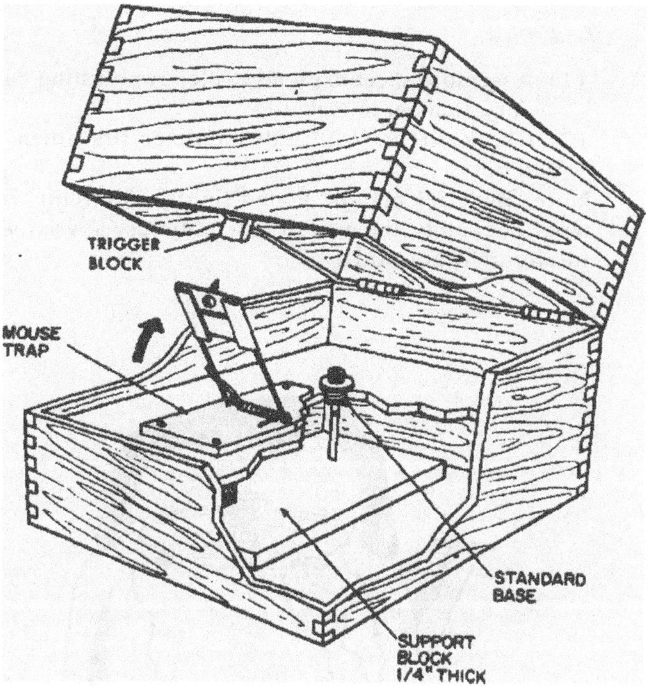

Electric Iron.

(1) Remove bottom plate.
(2) Insert bulk explosive and electric blasting cap.
(3) Attach shortened leg wires to power inlet.

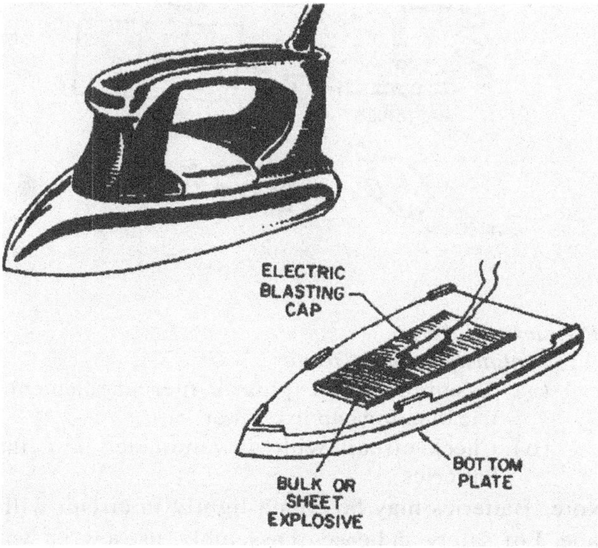

Teakettle.

(1) Assemble sheet explosive, electric blasting cap and mercury element in teakettle.

(2) Check circuit with galvanometer first, then install batteries.

Note. Batteries may be bound tightly in circuit with friction tape. For safety and ease of assembly, use a wrist watch delay in circuit (para 60d).

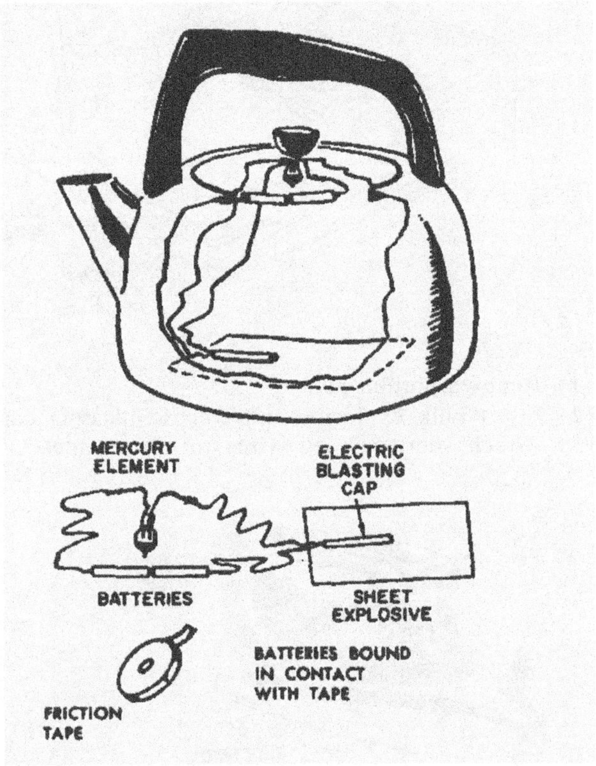

e. *Pressure Cooker.*

(1) *Antidisturbance circuit.*

(a) Assemble sheet explosive, mercury element, and electric blasting cap in cooker.

(b) Check circuit with galvanometer *first,* then install batteries.

Note. Batteries may be bound-tightly in circuit with friction tape. For safety and ease of assembly, use a wrist watch delay in circuit (para *60d).*

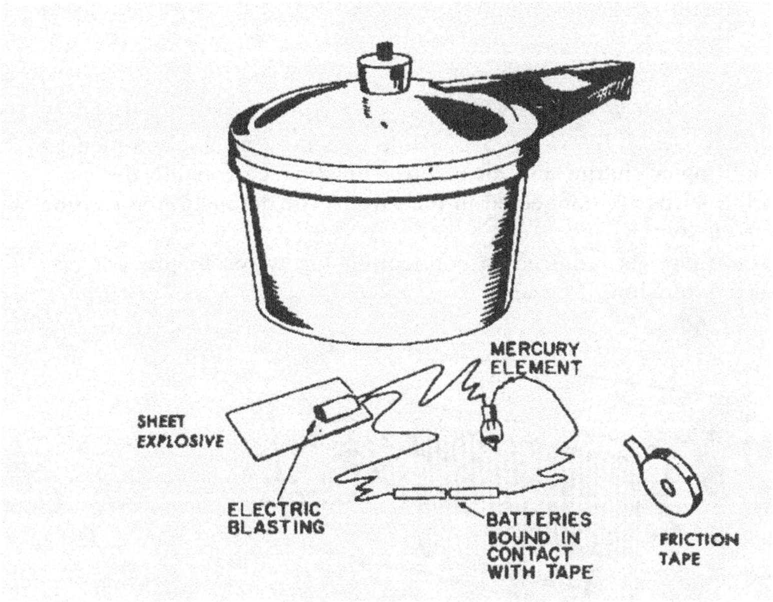

(2) *Loop switch.*

(1) Assemble sheet explosive and electric blasting cap.

(2) Cut leg wires to proper length. Remove insulation from ends and twist to form loop switch.

(3) *Check circuit with galvonometer.*

(4) Fasten one leg wire (insulated) to lid to serve as pull wire.

(5) Secure batteries in circuit by wrapping tightly with friction tape.

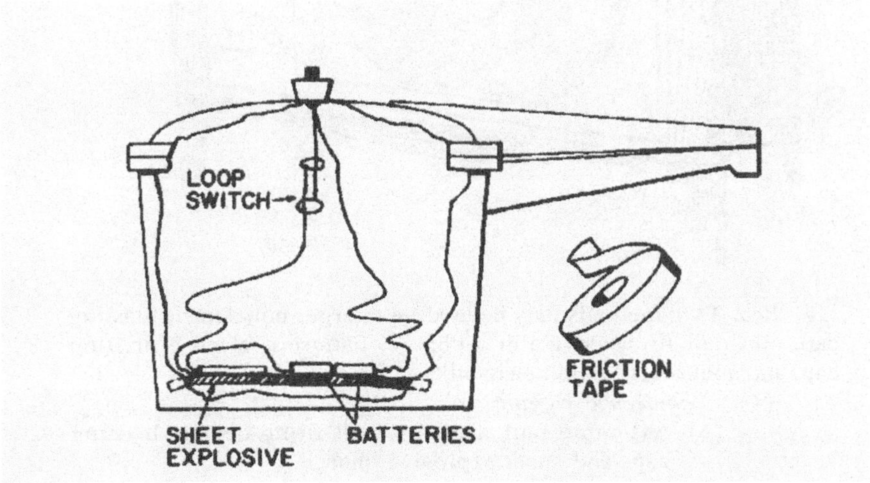

/. Radio and Television Sets. Both sets may be boobytraped by assembling a charge and an electric blasting cap inside the case. The leg wires are connected in the circuit for detonation at turning of off-on switch.

Extreme care is required in connecting leg wires to prevent premature explosion.

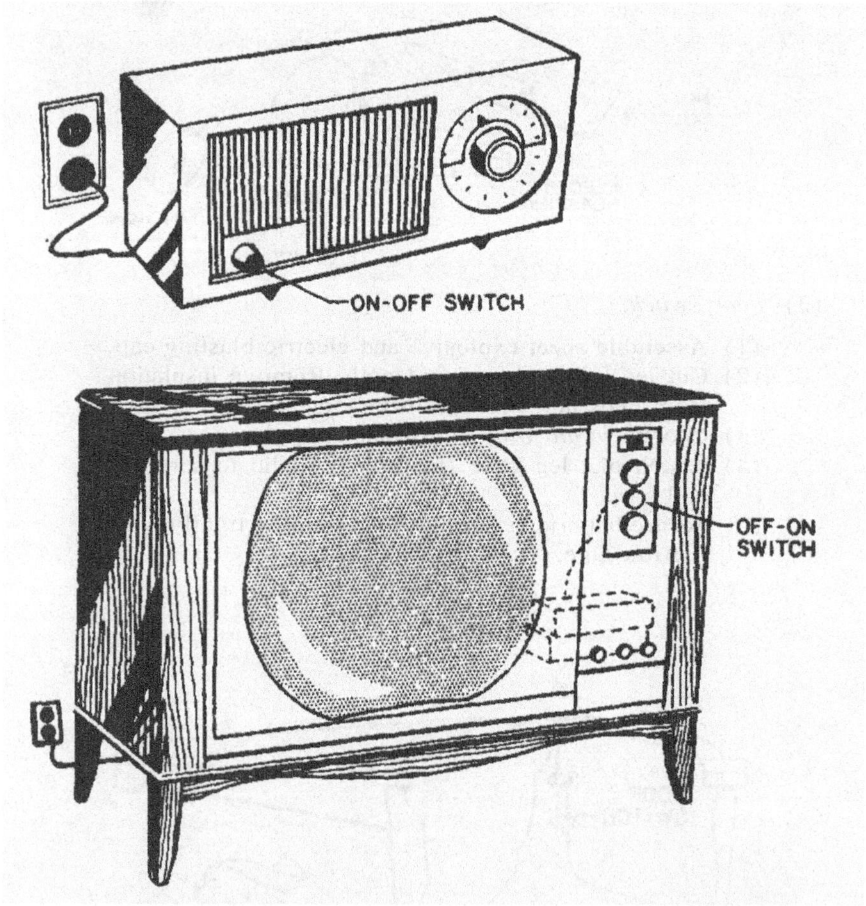

g. Bed. Two methods may be used—a charge, nonelectric blasting cap, and pull firing device or a charge, batteries, electric blasting cap, and a mercury switch element.

(1) *Nonelectric rigging.*

(a) Assemble pull wire, Ml pull firing device, blasting cap, and sheet explosive charge.

(b) Anchor pull wire so that a person sitting or lying on bed will initiate firing device.

(c) Conceal boobytrap.

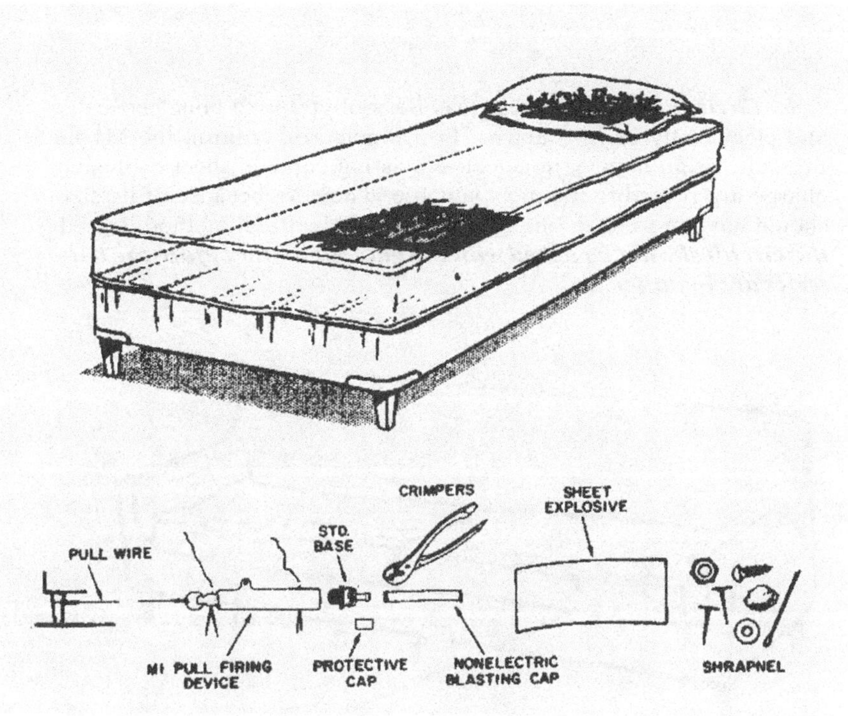

(2) *Electric rigging.*

(a) Assemble sheet explosive charge, electric blasting cap, and mercury element.

(b) Check circuit with galvanometer.

(c) Place boobytrap on bed to initiate when its level position is disturbed.

(d) Install batteries in circuit by wrapping tightly with friction tape.

(e) Conceal boobytrap.

Note. For safety and ease of assembly, use a wrist watch delay in circuit (para 60d).

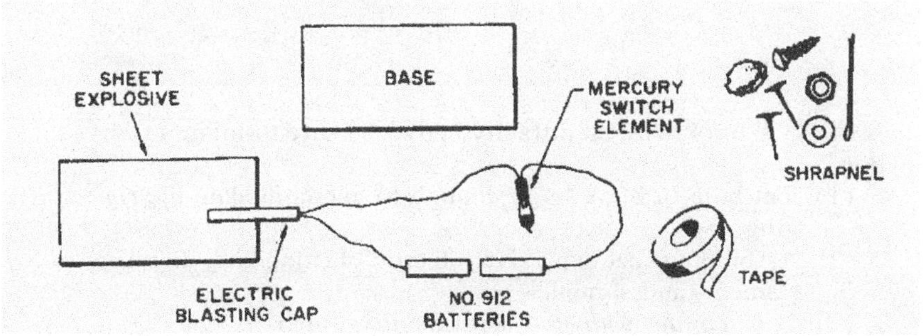

h. Chairs and Sofas. These may be boobytrapped nonelectrically and electrically as in *l* above. For nonelectric rigging the M1A1 pressure firing device, nonelectric blasting cap and sheet explosive charge are probably the most suitable. The sofa because of its size should have more than one rigging. If the electrical method is used ***the circuit should be tested with the galvanometer before the batteries are installed.***

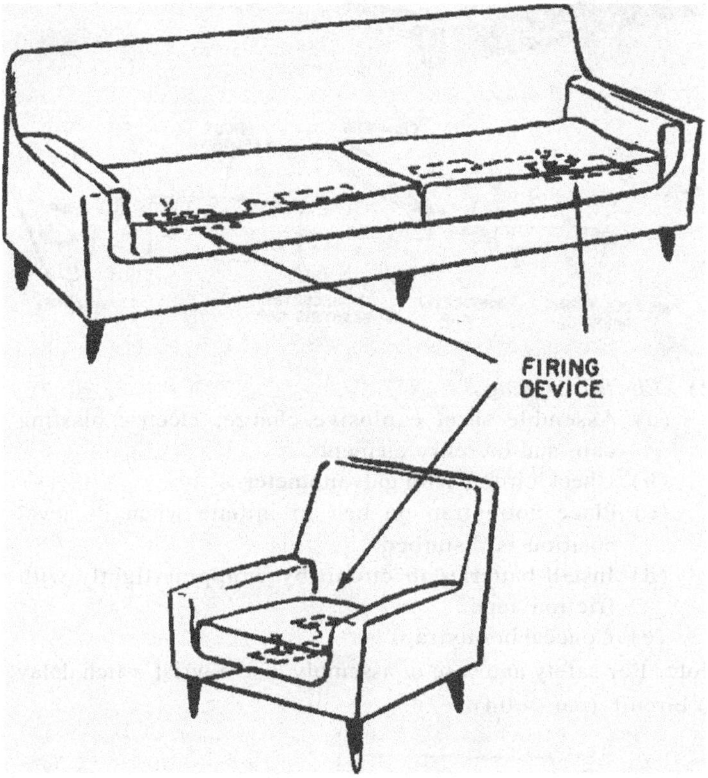

i. Book. A book with an attractive cover is sure to invite examination.

(1) Cut hole in book large enough to accommodate the rigging.
(2) Assemble sheet explosive, electric blasting cap, mercury element, and shrapnel.
(3) *Test circuit with galvanometer first,* then
(4) Secure batteries in circuit by wrapping tightly with friction tape.

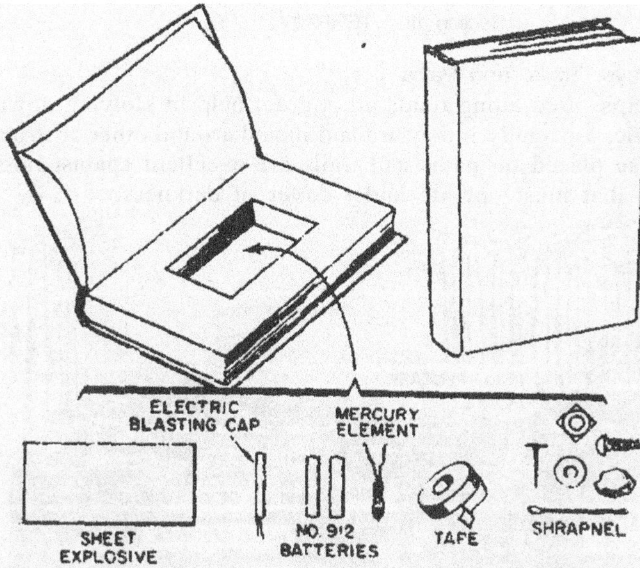
ELECTRIC
BLASTING CAP
MERCURY
ELEMENT
SHEET
EXPLOSIVE
NO. 9½
BATTERIES
TAPE
SHRAPNEL

45. Highways, Trails, and Paths

Boobytraps used along roads are a great help in slowing down enemy traffic, especially if they are laid in and around other obstructions. Those placed on paths and trails are excellent against raiding parties that must operate under cover of darkness.

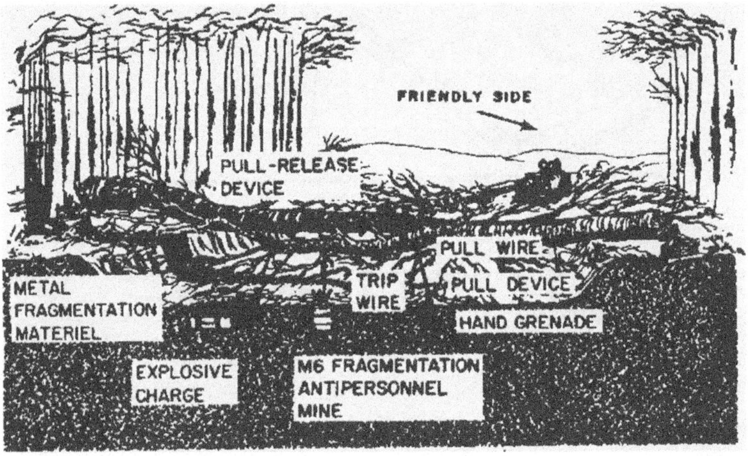

46. Locations

Boobytraps in roadway obstructions should be concealed on the enemy side. If the obstruction is heavy, requiring force to remove it, boobytraps concealed underneath will increase its effectiveness. Fragmentation charges are very destructive against personnel. These include hand grenades; bounding antipersonnel mines with their own special fuzes actuated by pressure or trip wire; ordinary explosive charges covered with pieces of scrap metal, nails, gravel, lengths of wire, nuts and bolts; and the like. The latter may be actuated by any of the standard firing devices—by pressure, pressure-release, pull-release, and pull.

a. The jet of the M2A3 shaped charge from the roadside directed into a moving vehicle is very destructive.

(1) Assemble an M3 pull-release firing device and detonator, ***length of detonating cord, priming adapter, and non-electric*** blasting cap.

(2) Drive anchor stake in berm at side of road and attach pull wire. Drive stake or lay log, stone, or other object on other side to support pull wire at proper height off ground.

(3) Attach firing device assembly to stake at proper position.

(4) Fix shaped charge in position to direct explosive jet into vehicle when front wheels hit trip wire.

(5) Attach free end of pull wire in hole in winch and draw taut.

(6) Screw priming adapter'and nonelectric blasting cap in threaded cap well.

(7) Conceal boobytrap.

(8) Arm firing device.

Note: Cone may be filled with fragments.

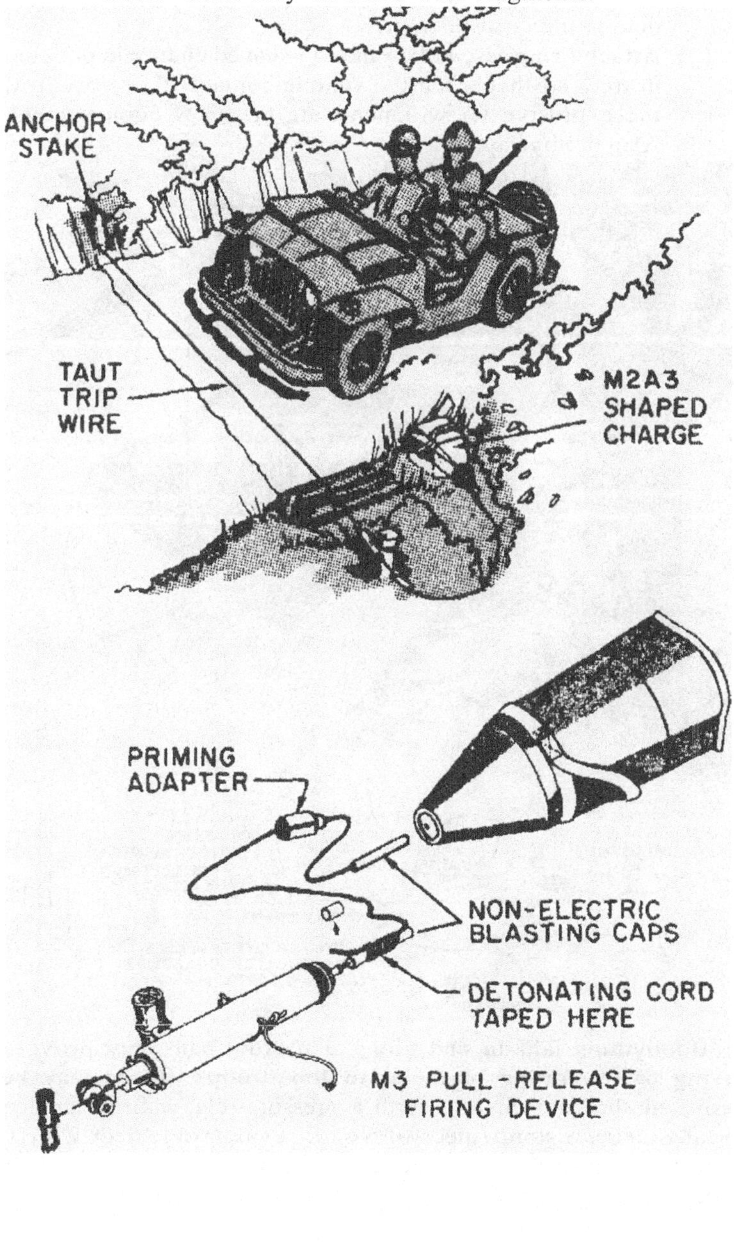

b. An M3 shaped charge boobytrap placed overhead in a tree in a wooded area will destroy both tank and crew if located properly. Trip wire, being very thin and camoufloage-colored, is not easily detected by a driver.

(1) Assemble two firing devices (only one may be necessary) with detonators and lengths of detonating cord and a detonating cord primer.

(2) Attach firing assemblies and M3 shaped charge in position in tree, so that when the vehicle contacts the trip wires, the explosive jet will penetrate the crew compartment.

(3) Arm boobytrap.

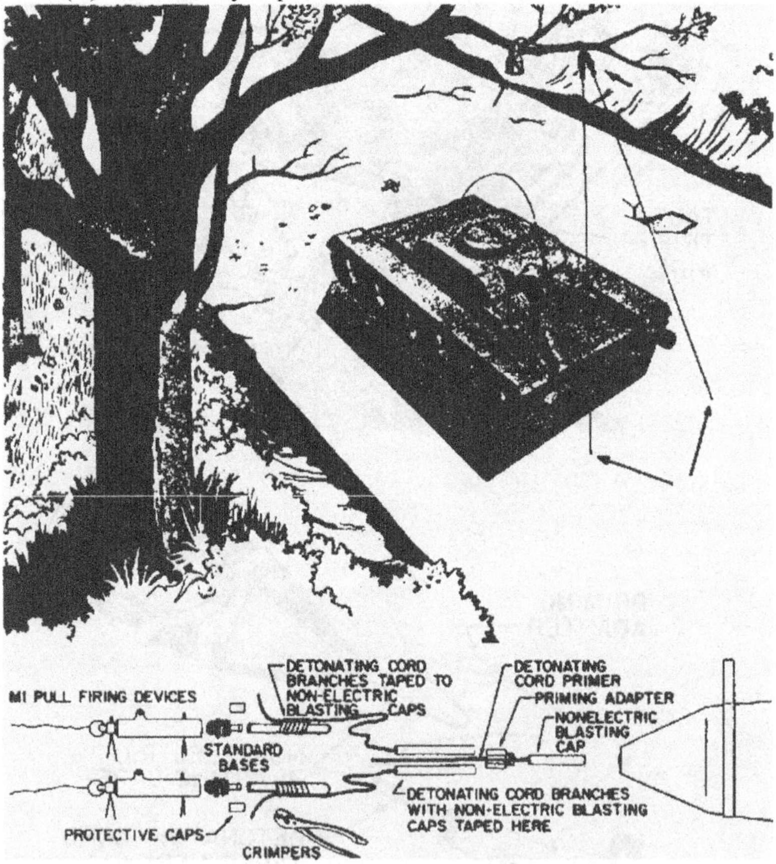

c. Boobytraps laid in and along a narrow path may prove a delaying or frustrating obstacles to foot troops. These may be improvised shrapnel charges with a pressure-release firing device concealed under a stone, piece of wood, or other object, or with a pull or pull-release firing device and a trip wire. The latter would be very effective against patrols.

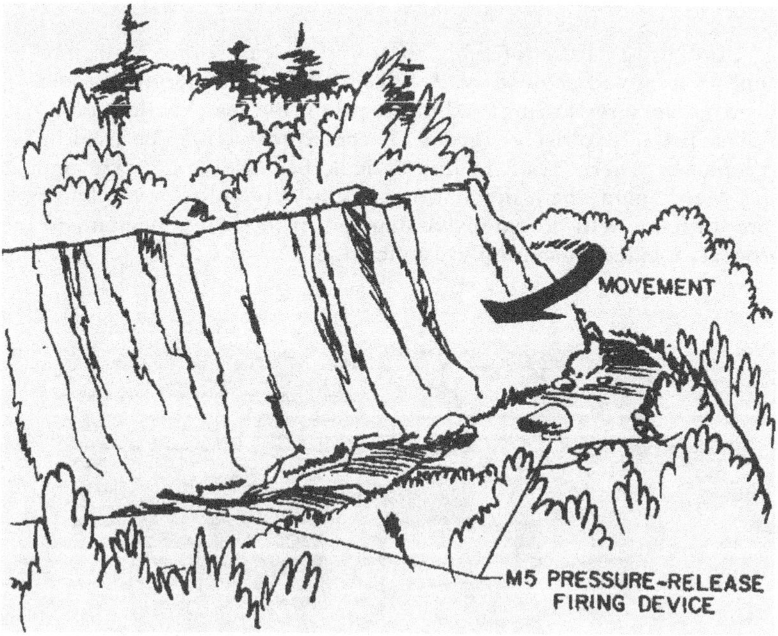

47. Special Locations

a. Abandoned serviceable or repairable items are frequently boobytrapped if time and equipment are available. Even unserviceable items may be rigged against scavangers who may search through the wreckage for useful things.

b. Abandoned ammunition should be exploited to the maximum. Chain detonations of connected mines or sections of bangalore torpedo are particularly effective.

c. Boobytraps are applicable to storage areas where materials cannot be removed or destroyed. Several charges strategically laid will prove very rewarding. A lumber pile provides excellent concealment for an explosive rigging. Sheet explosive may be used in many places where TNT is impractical, because of its size and shape. Here again chain detonations of explosive blocks and bangalore torpedos will do extensive damage, if the firing mechanism is properly located and cunningly concealed.

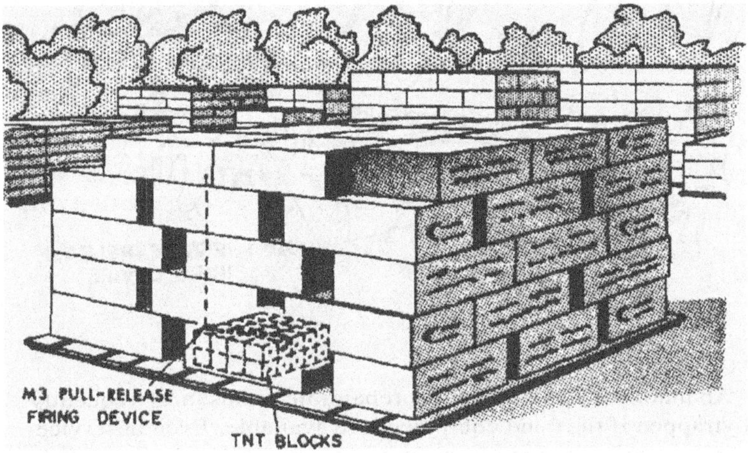

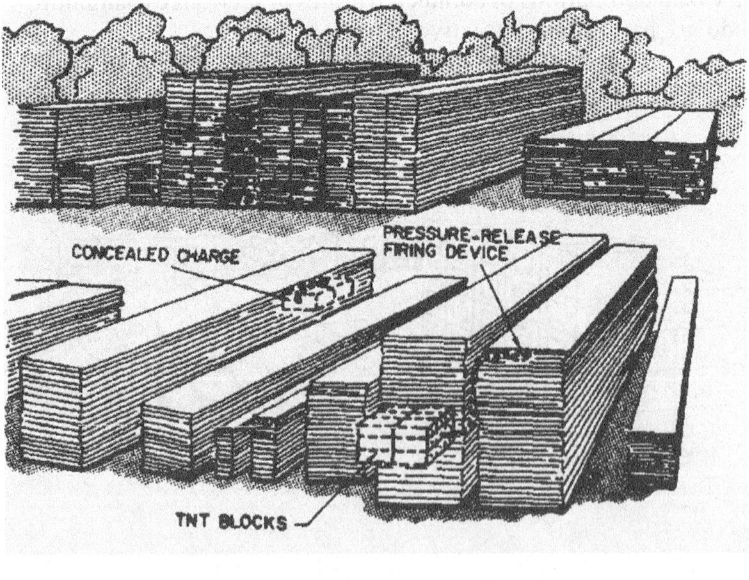

43. Abandoned Vehicles

a. Truck Wheel.

(1) Insert length of heavy wire in interceptor hole in firing device.

(2) Remove safety pin and replace with length of thin wire. Bend both wires slightly to prevent falling out.

(3) Assemble standard base, nonelectric blasting cap, and firing device.

(4) Assemble two 2-block explosive charges, nonelectric blasting caps, priming adapters, and length of detonating cord.

(5) In hole prepared under truck wheel, assemble bearing blocks (take weight off explosive charge), charges, bearing board, protective blocks (take weight off firing device), and firing device.

(6) Arm firing device.

(7) Cover boobytrap, and camoulflage.

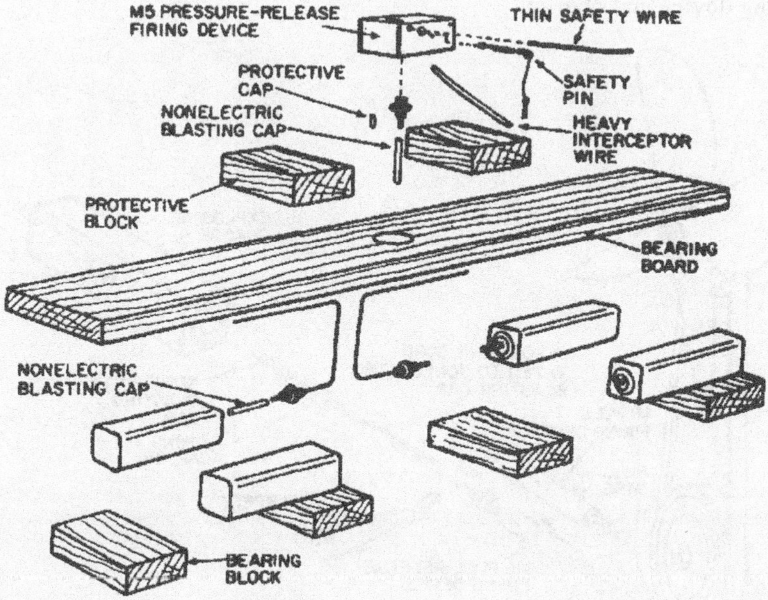

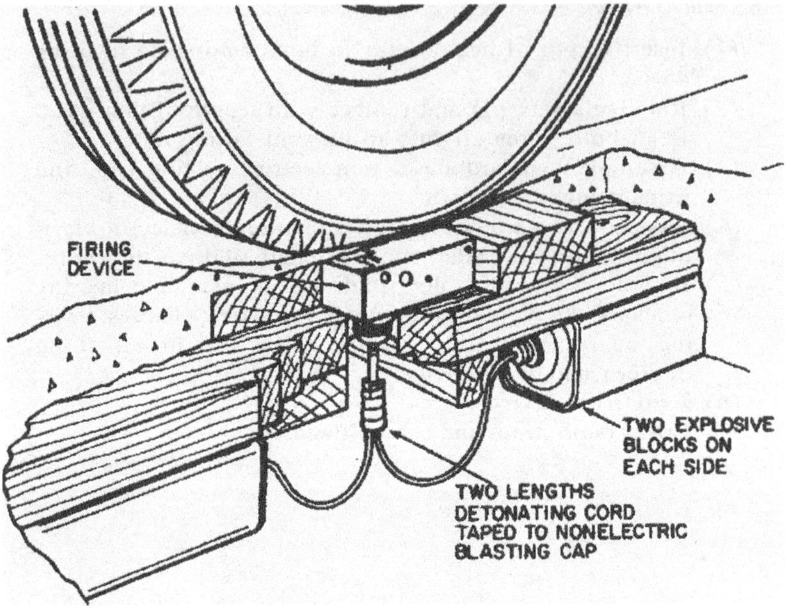

b. Motor. The fan belt is an excellent anchor for a pull wire. The pull wire will be much harder to detect if anchored underneath the bottom pulley, from where it may be extended any length to the firing device and charge.

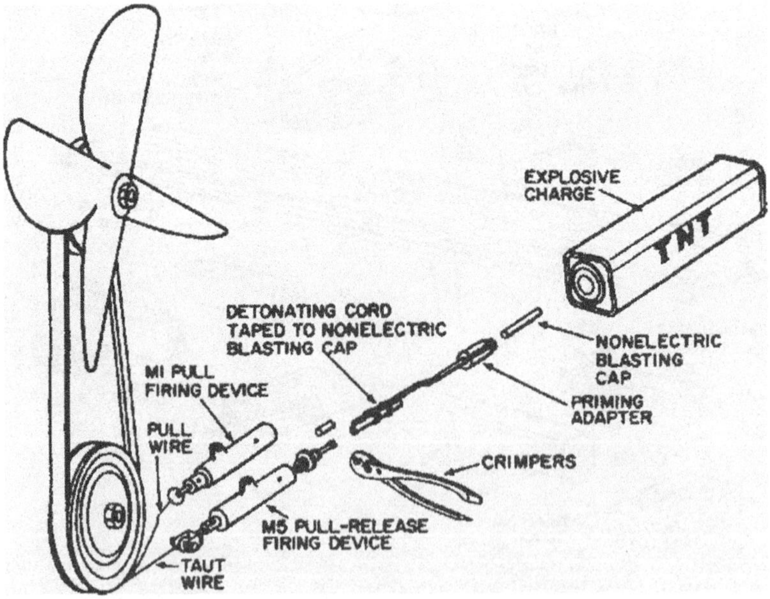

c. Electric System. A useful combination is a charge primed with an electric blasting cap with clamps attached to the leg wires. This may be attached to detonate by turning on the ignition switch, engaging the starter, braking, and the like.

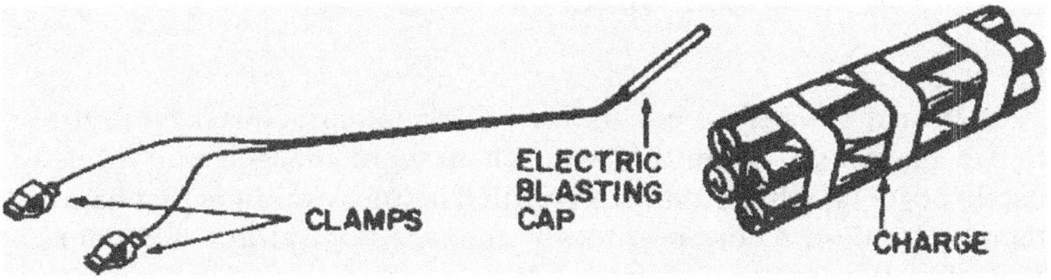

d. Body. Another combination useful in rigging a seat or any other part of the vehicle body is a charge detonated electrically by means of a mercury switch element.

(1) Assemble charge, electric blasting cap, and mercury element.

(2) Place boobytrap in position and check circuit with a galvanometer.

(3) Attach batteries in circuit by wrapping tightly with friction tape.

Note. Always check circuit before attaching batteries.

This rigging may be assembled in a small package for use in a seat cushion or separated for convenience for another location in the body of the vehicle.

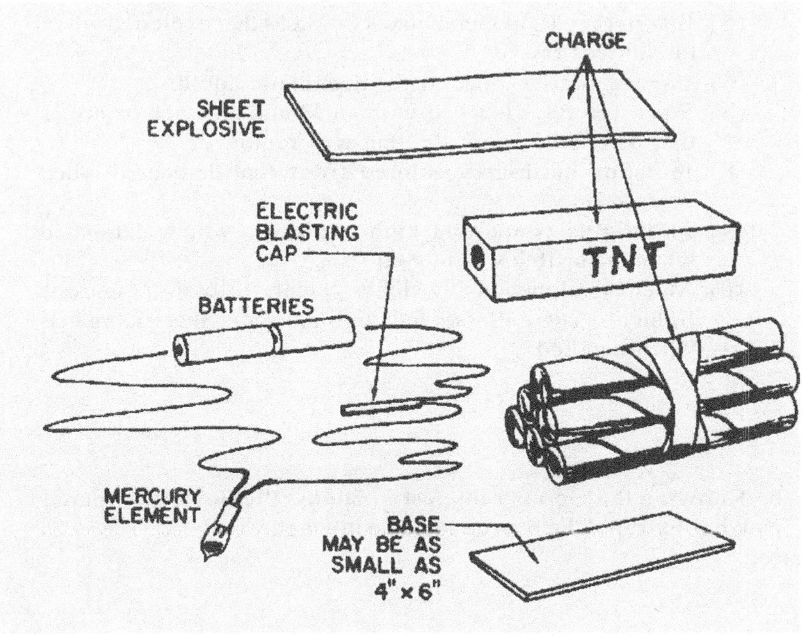

CHAPTER 5

MISCELLANEOUS BOOBYTRAPS

Section I. STANDARD BOOBYTRAPS

49. Tactical Use

In World War II, every major power manufactured boobytraps to use against the enemy. Most of them were charged imitations of useful objects, which maimed or killed helpless soldiers that handled them. The defect common to all standard boobytraps however, is that after the first or second explosion, all others of the same type become ineffective. A "one-shot" job hardly justifies production costs.

50. Foreign Types

a. The Soviets used more standard boobytraps in World War II than any other combatant. A weird assortment of charged imitations of items issued to German soldiers were dropped from Soviet planes. Some of these were:

(1) Cartridge boxes, apparently filled with ammunition, containing high explosives and detonators.
(2) Bandage packets containing detonators and shrapnel.
(3) Bandage cases with Red Cross insignia rigged as mines.
(4) Rubber balls, about twice the size of a fist that detonated upon impact.
(5) Silver-grey light metal boxes or Masks that exploded when the lid was raised.
(6) Cognac bottles filled with incendiary liquid.
(7) Small red flags marked with an M and attached to mines that detonated when the flag was removed.
(8) Imitation earth-grey colored frogs that detonated when pressed on.
(9) Flashlights containing high explosive which detonated when the switch was moved.
(10) Mechanical pencils, watches, cigarette cases, cigarette lighters, salt cellars, and similar items that detonated when handled.

b. Knowing the German interest in books, the Soviets prepared a book boobytrap. The charge inside detonated when the cover was raised.

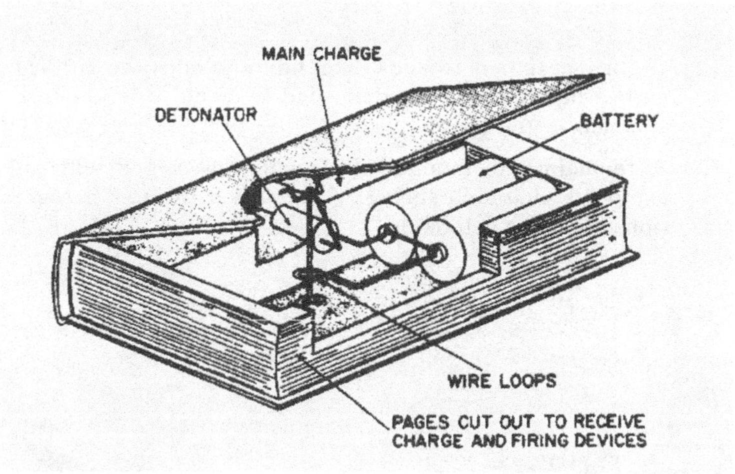

c. The British also had a book boobytrap; but it was slightly more complicated than the Soviet version, above.

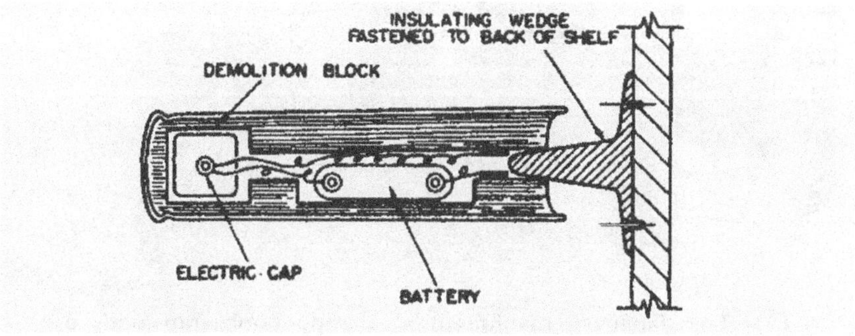

d. All sorts of dirty-trick devices were used by the enemy.

(1) A flashlight was rigged with a charge and an electric detonator powered and actuated by the original dry cell battery switch, and circuit.

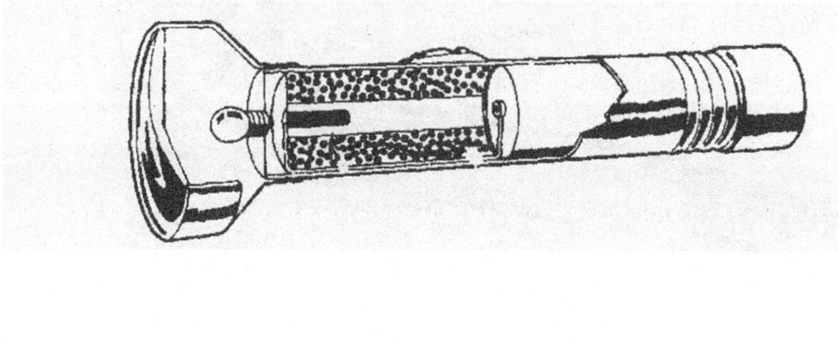

(2) Bottles designed to look like liquor bottles were filled with a liquid explosive detonated by a pull-friction fuze attached to the cork.

(3) A fountain pen, though very small, was rigged with an explosive charge, a spring driven striker to fire a percussion cap, and a detonator.

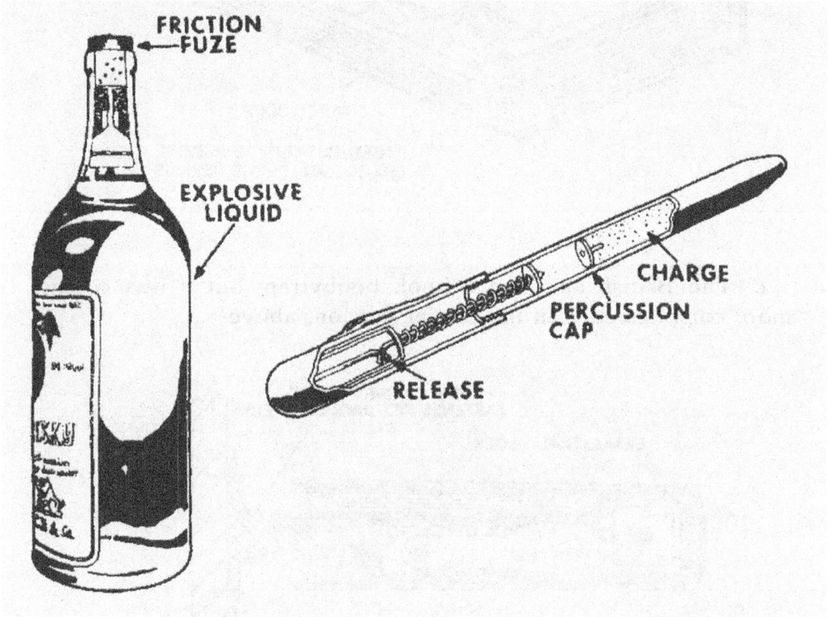

(4) The Japanese manufactured a pipe boobytrap with a charge, detonator, and spring-loaded striker.

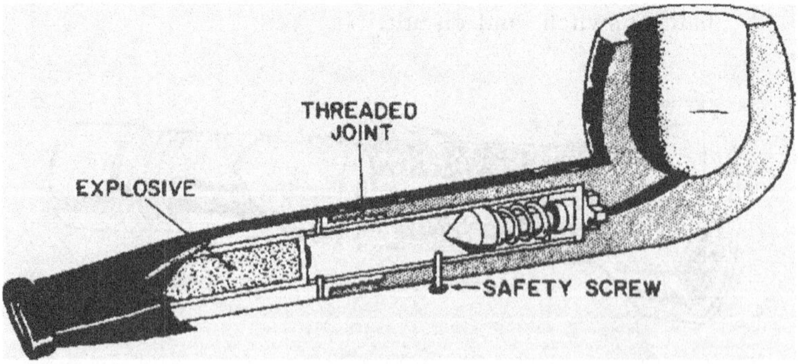

(5) The Italians had a boobytrapped headset containing an electric detonator connected to the terminals on the back. The connection of the headset into the live communication line initiated detonation.

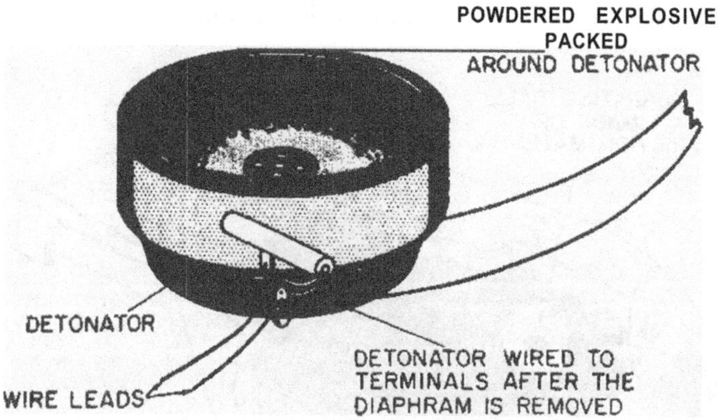

(6) The Germans converted their own and enemy standard canteens into boobytraps. The explosive charge was detonated by a pull fuze and a pull wire connected to the cap. When partially filled with water and placed in its canvas case, it was very deceptive. The canteen boobytrap had an effective radius of 3 to 5 yards.

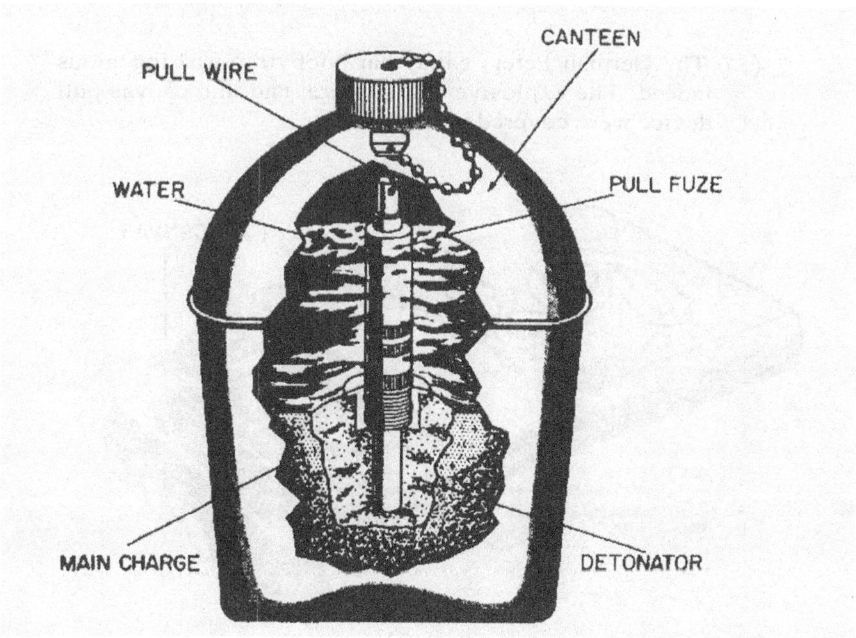

(7) Another German device was the boobytrap whistle. This consisted of a policeman's or referee's whistle with a charge and a metal ball covered with a layer of friction compound. Blowing the whistle moved the ball, igniting the friction compound and detonating the charge.

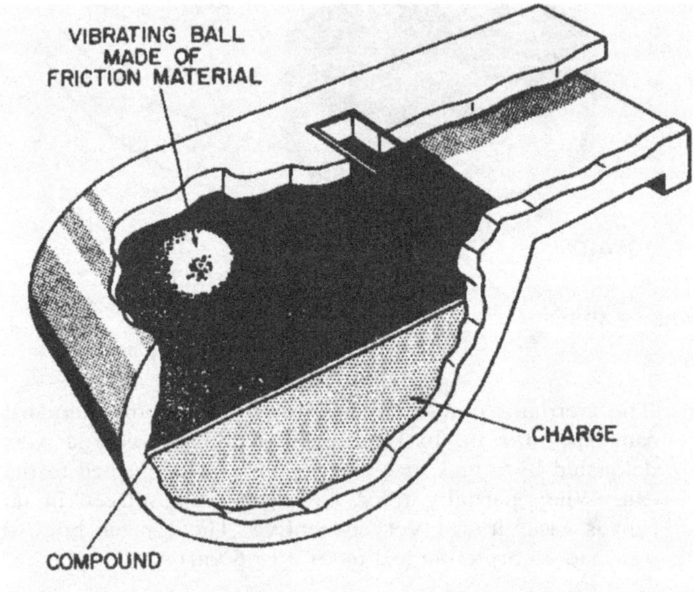

(8) The German Peters candy bar boobytrap was ingenious indeed. The explosive charge, fuze, and thin canvas pull device were covered with chocolate.

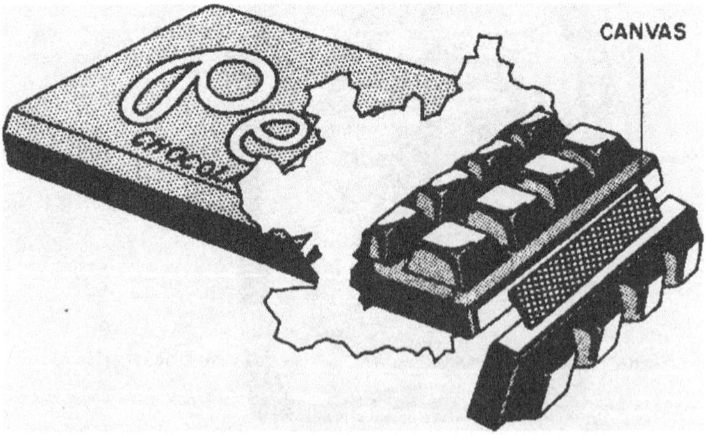

Section II. IMPROVISATIONS

51. Ingenuity

a. Through information on military operations in World War II, the U.S. soldier has been well-prepared for the dangerous mission of laying, detecting, and disarming boobytraps in conventional warfare. However, he now is virtually a novice in comparison with the cunning and ingenious present day guerrilla, who at the start was almost totally lacking in material and equipment.

b. Experience has shown that in guerrilla warfare, carried on by illy-equipped native populations, boobytrapping success depends largely on ingenuity. Explosive, a necessary element, is either improvised from commercial ingredients or captured from the enemy. Captured mines, ammunition, and other similar material are disassembled and every ounce of explosive saved.

52. Training

Every soldier should have some training in the lessons learned from the guerrillas, for many items they have improvised and the way they have used them are also applicable to conventional warfare. With little effort, a soldier may be trained so that with no military equipment whatever but with ample funds, he may prepare himself to fight effectively with materials available from merchants, junk piles, and salvage.

53. Application

The improvisations included in this section are gathered from numerous sources. Some may have wider application to boobytrapping than others. How the guerrilla may use them, however, is unpredictable. All are presented to stimulate initiative and arouse enthusiasm to out-do backward enemy peoples in devising and placing boobytraps and to develop a higher level of proficiency than ever before in their detection and removal.

54. Improvised Time Fuze and Explosive Caps

a. Fast burning fuse (UO inches per minute).

(1) Braid three lengths of cotton string together.

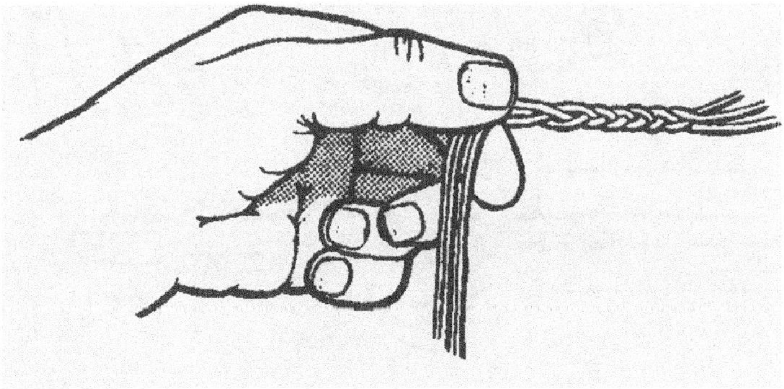

(2) Moisten fine black powder to form a paste. Rub paste into twisted string with fingers and allow to dry. If a powder is not available, mix 25 parts potassium nitrate (saltpeter) in an equal amount of water and add 3 parts pulverized charcoal and 2 parts pulverized sulphur to form a paste. Rub paste into twisted string and allow to dry.

(8) Check burning rate before using.

b. Slow burning fuse (2 inches per minute).

(1) Wash three lengths of string or three shoelaces in hot soapy water and rinse.
(2) Dissolve 1 part potassium nitrate or potassium chlorate and 1 part granulated sugar in 2 parts hot water.
(3) Soak string or shoelaces in solution and braid three strands together. Allow to dry.
(4) Check burning rate.
(5) Before using, coat several inches of the end to be inserted into cap or material to be ignited with black powder paste (*a* (2) above).

c. Electric Blasting Cap.

(1) With file or other instrument make hole in end of light bulb.
(2) If jacket is not available, solder or securely fasten two wires to bulb—one on metal threads at side and other at metal contact on bottom.
(3) Fill bulb and empty portion of blasting cap with black powder. Tape blasting cap on top of bulb.

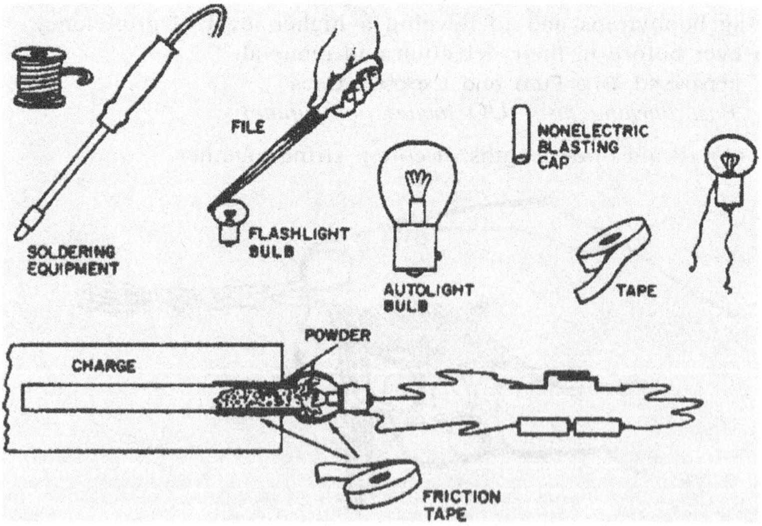

d. Percussion Cap Assembly.

(1) Remove projectile, but not powder, from small arms cartridge.

(2) Tape nonelectric blasting cap securely in cartridge.

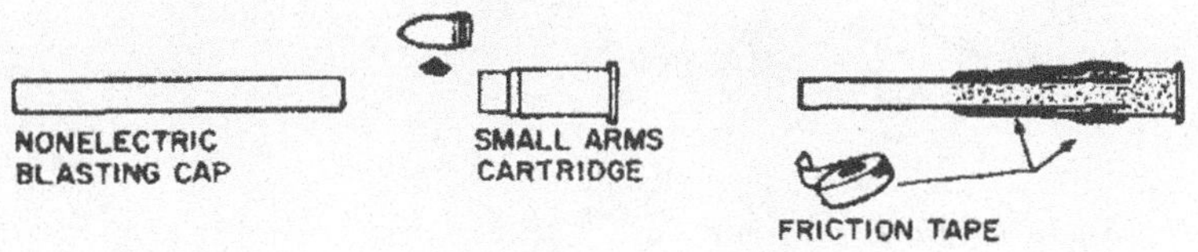

55. Pull Firing Devices

a. Tube and Striker.

Assemble tube, spring, striker shaft with hole or with hex nut, soft wood or metal top plug, pull pin, and improvised percussion cap assembly.

Note. Always assemble firing device before attaching the improvised percussion cap assembly.

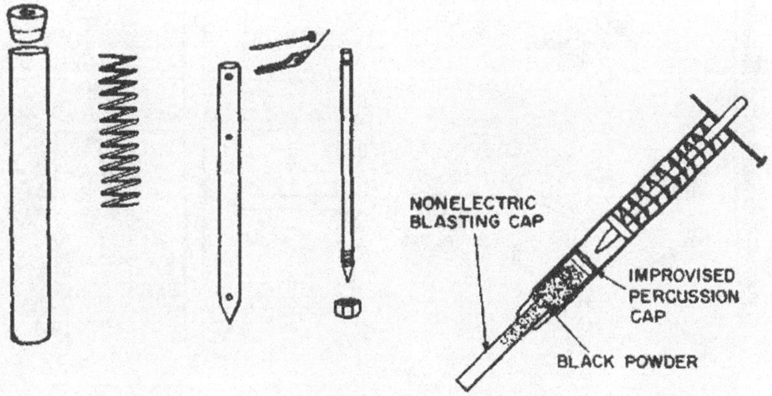

b. Clothes Pin.

(1) Wrap stripped ends of leg wires around clothes pin jaws to make electrical contact.

(2) Assemble charge, adapter, electric blasting cap, and clothes pin.

(3) Insert wooden wedge, anchor clothes pin, and install trip wire.

(4) Check circuit with galvonometer *first,* then connect batteries.

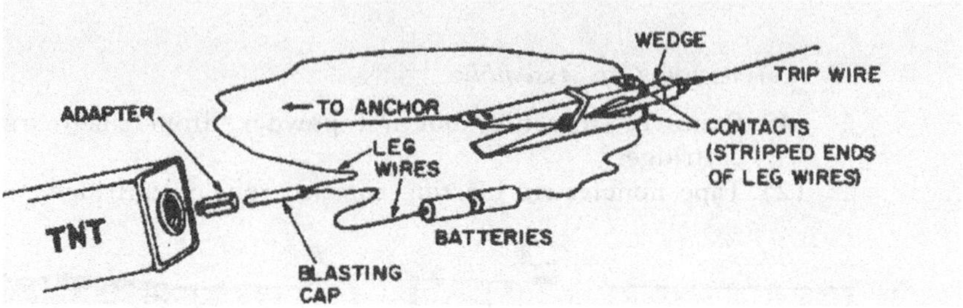

c. *Stake or Pole Initiator.*

(1) Assemble stake or pole, container, metal contact plates, charge, electric blasting cap, and pull cord.

(2) Check circuit with galvonometer *first,* then connect batteries.

(3) Fasten down top of container and seal hole around stake with friction tape.

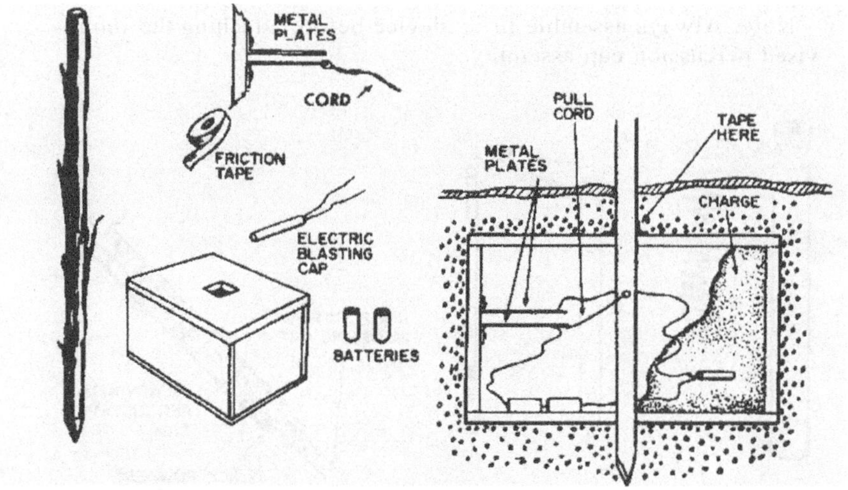

d. *Rope and Cylinder.*

(1) Cut leg wires to proper length.

(2) Prepare wooden end plugs and bore hole in one to receive leg wires.

(3) Thread leg wires through hole in block.

(4) Strip end of one leg wire and twist into loop, and secure other leg wire in position.

(5) Test circuit with galvonometer.

(6) Assemble metal cylinder, contact bolt, pull cord, charge, blasting cap, end blocks, and batteries.

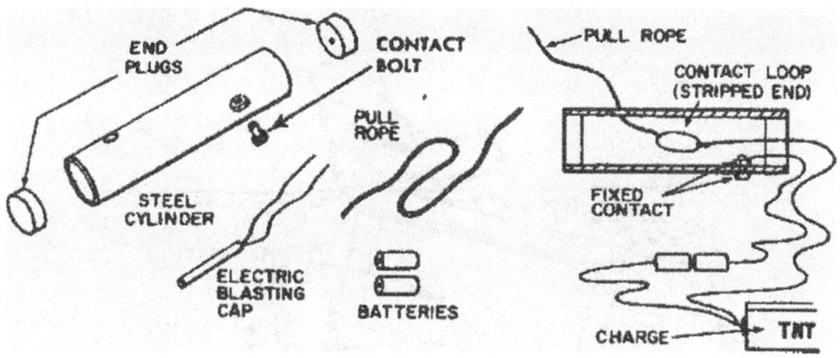

e. Trip Lever and Putt Pin.

(1) *Flat placement.*
Assemble container, charge, improvised pull firing device (o above) and trip lever.

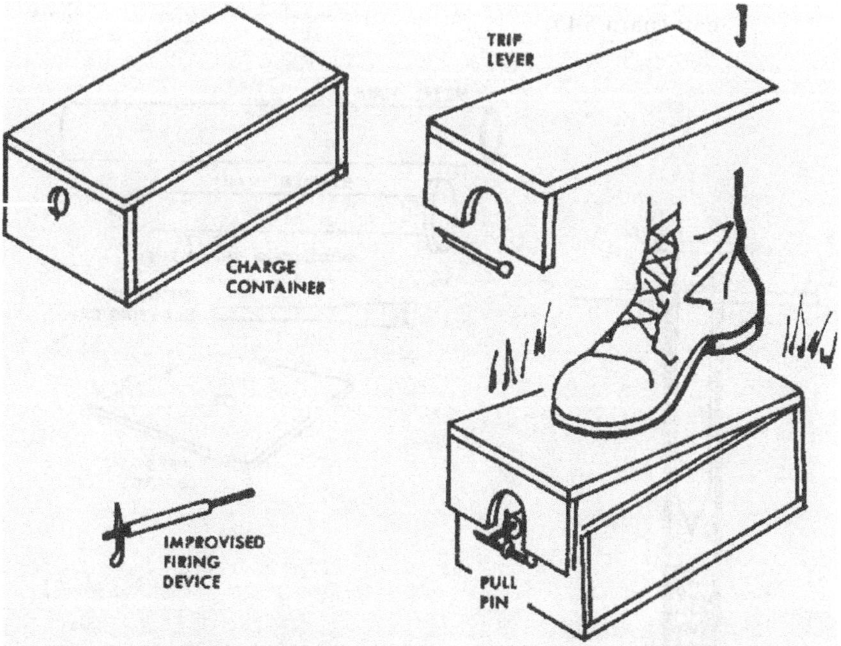

(2) *Sloping placement.*
Assemble container, charge, improvised firing device (*a* above) and stake.

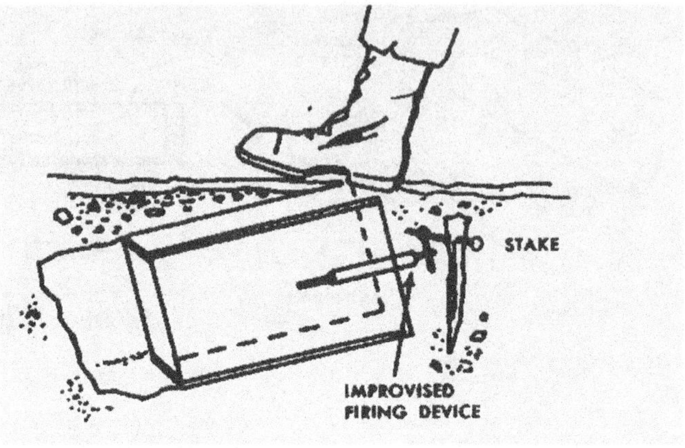

56. Pressure Firing, Devices

a. Mechanical Concussion.

(1) Force striker into hole in pressure board.

(2) Insert wood or soft metal shear pin in shear pin hole.

(3) Assemble striker, metal tube, and improvised blasting cap (para 54).

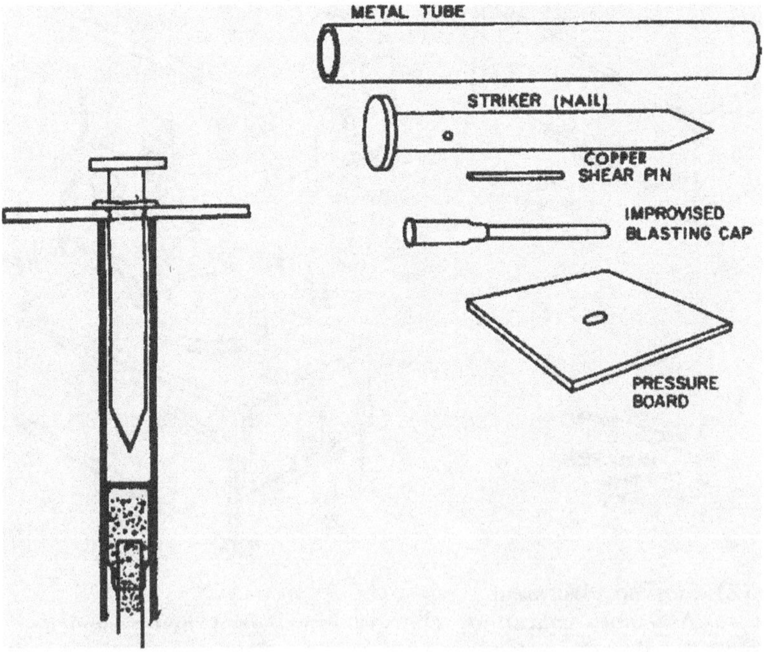

6. *Electrical.*

(1) *Lever arm.*

(a) Attach contact blocks to ends of wooden levers.
(b) Assemble wooden levers, rubber strip, and plastic sponge.
(c) Attach leg wire contacts.

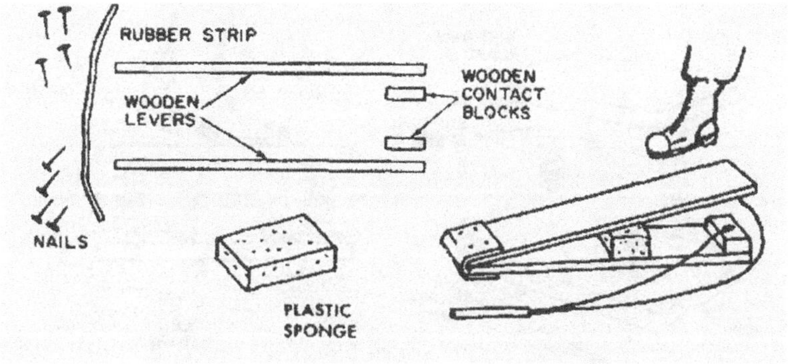

(2) *Flexible side.*

(a) Attach metal contact plates to bearing boards.
(b) Thread leg wires through holes in lower bearing board and attach to contact plates.
(c) Attach flexible sides.

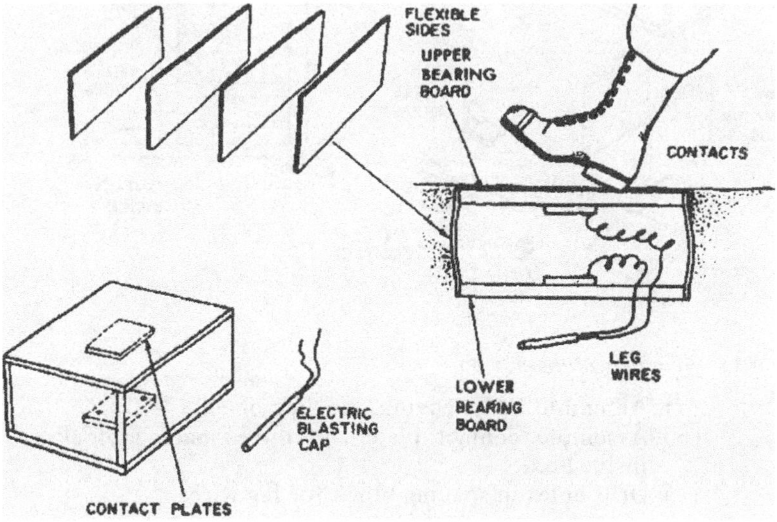

(8) ***Springed pressure board.***

(a) Assemble metal contacts, springs, bearing board, and'pressure board.

(b) Attach leg wires to metal contacts.

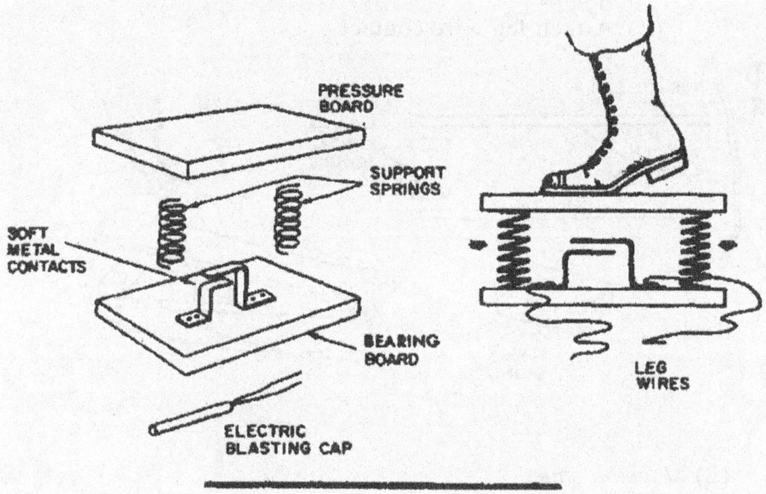

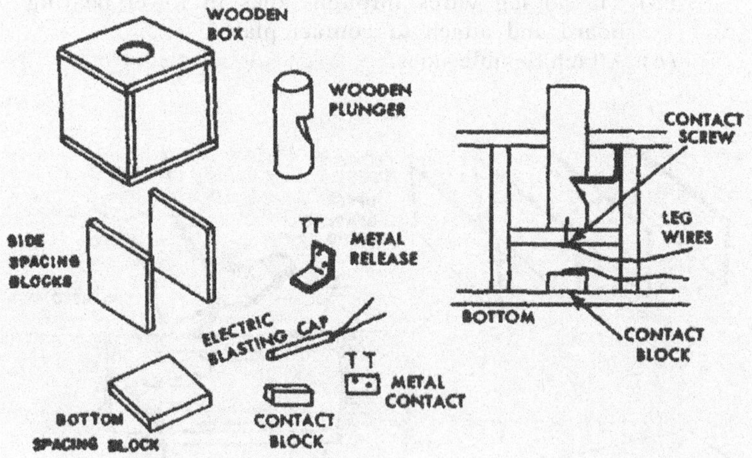

(4) *Wooden plunger.*

(a) Assemble box, leaving one side open.

(b) Assemble contact plate and three spacing blocks inside box.

(c) Drill holes in spacing block for leg wires.

(d) Assemble plunger, metal release, contact block, metal contact, and contact screw.
(e) Thread leg wire through holes in spacing block and attach to contacts.

(5) *Metal box.*

(a) Attach metal contact to wooden contact block.
(b) Assemble contact block and metal contact, brackets, metal release, plunger, and wooden box lid.
(c) Bore hole in side of box for leg wires.
(d) Thread leg wires through hole in box.
(e) Attach one leg wire to plunger, the other to metal contact.

Note. Batteries may be placed inside box if necessary.

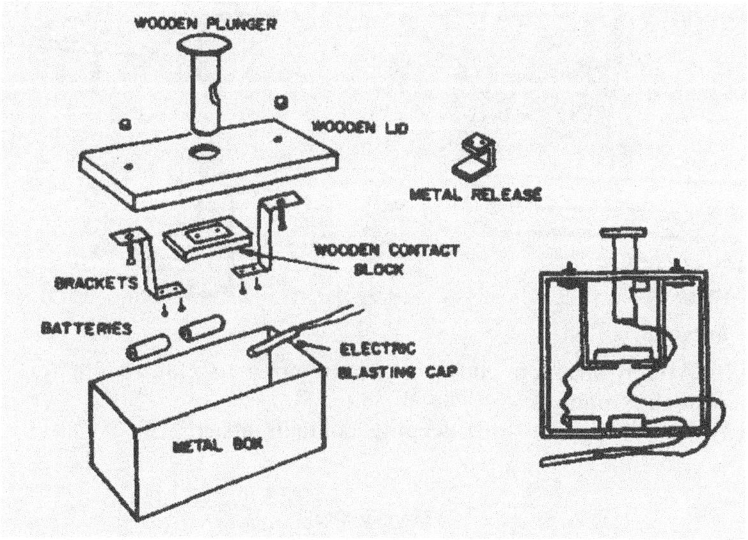

57. Tension-Release Bring Device

Attach stripped ends of circuit wires to ends of clothes pin to form contacts. Attach taut trip wires below contacts.

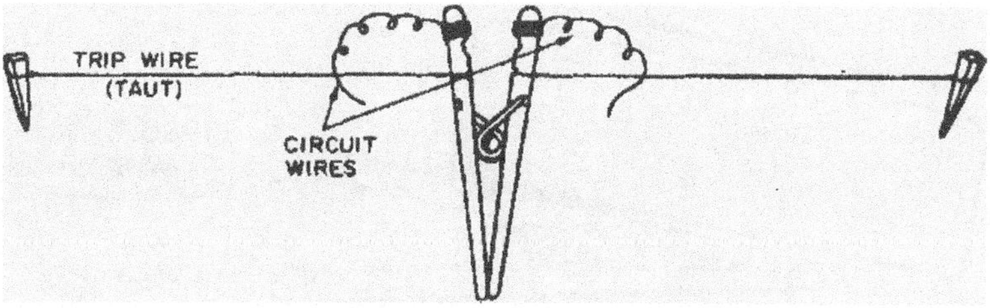

58. rVcawrt-RcJMM

a. Double Contact.

(1) Bore holes in top of mine body to accommodate long contacts.

(2) Assemble pressure board, coil springs, wooden contact board and metal contacts.

(3) Attach circuit wires.

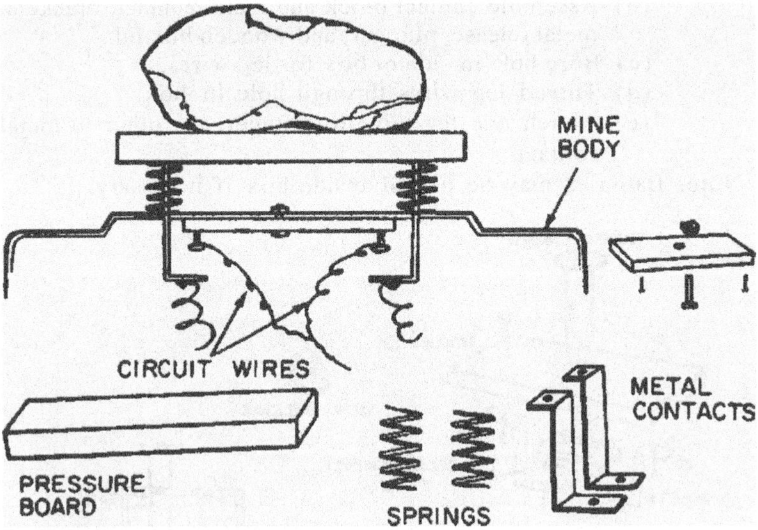

b. Clothes Pin.

(1) Attach stripped ends of circuit wires to clothes pin to make contacts.

(2) Place mine on top, keeping contacts apart.

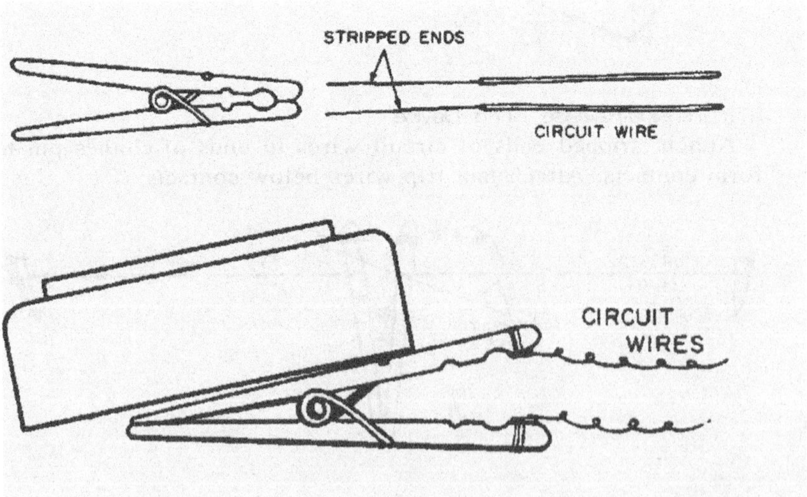

c. Bottom Plunger.

(1) Bore hole in bottom of nune case to admit plunger.
(2) Attach lower metal contact over hole.
(3) Assemble mine, pressure block, upper metal contact, and nonmetallic plunger.
(4) Attach circuit wires.

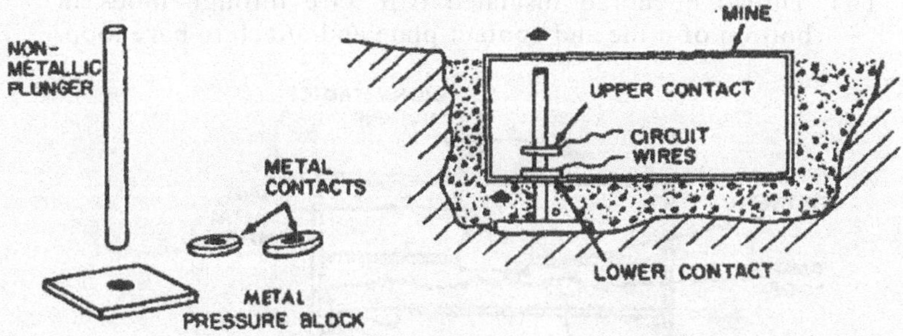

d. Mousetrap.

(1) *Mechanical*
See para 44 *b* (2)
(2) *Electrical*
(a) Remove triggering devices from mousetrap.
(b) Assemble trap, contact plate, and circuit wires.
(c) Place weight on top with striker in armed position.

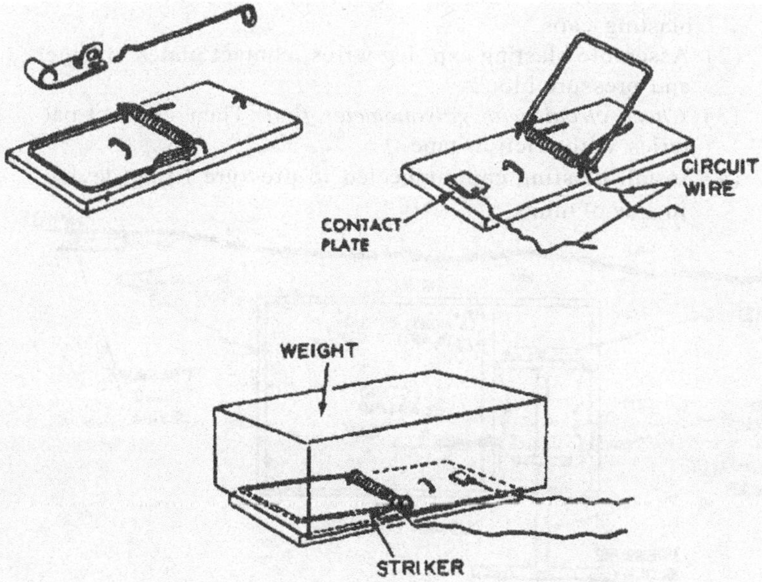

59. Anti-Lift Devices

a. Loop Contact.

(1) Drill hole in bottom of mine to admit insulated pull wire.

(2) Assemble plunger, metal release, and contact plate.

(S) Attach circuit wires and bare loop to plunger contact and contact plate.

(4) Thread anchored insulated trip wire through holes in bottom of mine and contact plate and attach to bare loop.

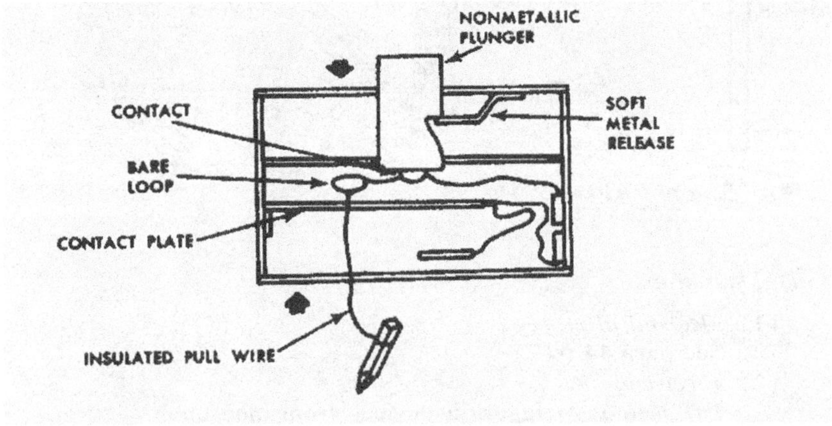

b. Double Detonator.

(1) Drill three holes—one in bottom, one in partition, and one in side—to admit nonmetallic plunger and two electric blasting caps.

(2) Assemble blasting cap, leg wires, contact plates, plunger and pressure block.

(3) *Check circuit with galvonometer first.* Then connect batteries with friction tape.

(4) Install blasting cap connected to pressure firing device in side of mine.

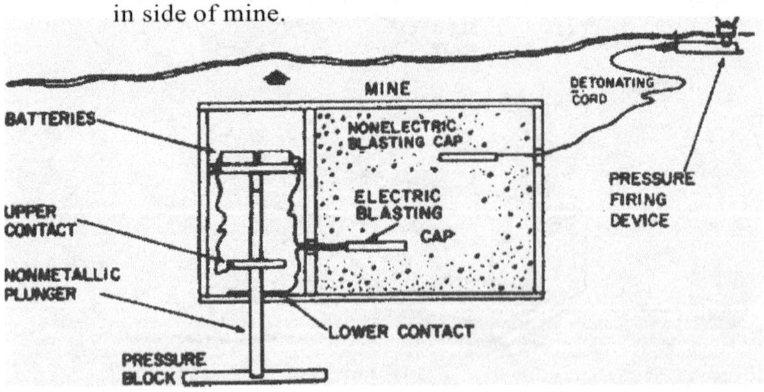

c. Sliding Contact.

(1) Assemble metal cap, nonmetallic tube or carton, sliding contact, wooden plug, and leg wires at contacts.

(2) *Check circuit with a galvonometer first,* then connect batteries with friction tape.

(3) Install assembly in tube.

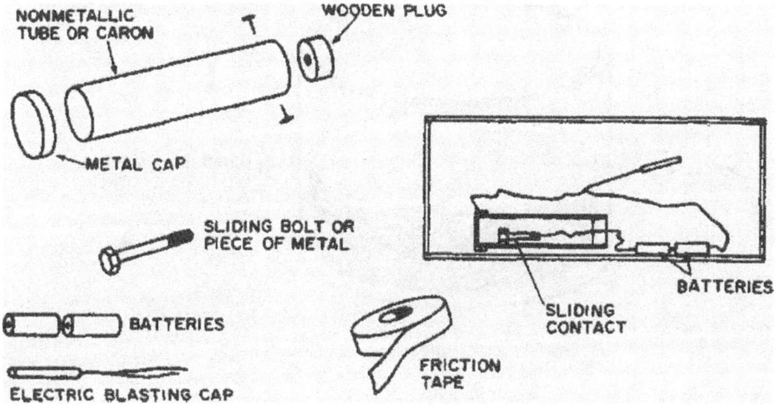

60. Delay firing Devices

a. Cigarette Timer.

(1) Test burning rate of time fuze and cigarette. (A cigarette usually burns at the rate of 1 inch in 7 to 8 minutes.)

(2) Cut sloping end on length of time fuze.

(3) Assemble sloped end of time fuze, match head, and cigarette.

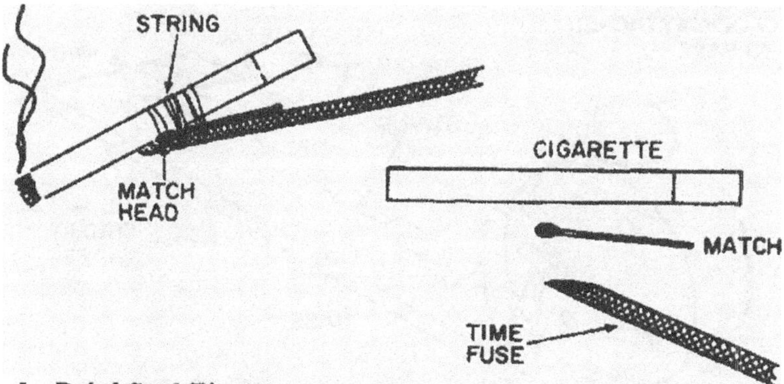

b. Dried Seed Timer.

(1) Determine expansion rate of seeds.

(2) Place in jar and add water.

(3) Assemble jar, lid, circuit wires, metal contacts, and metal disk and secure with friction tape.

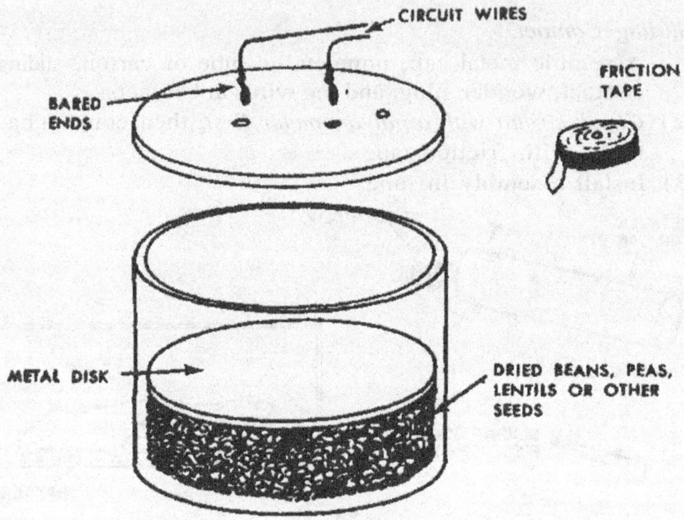

c. *Alarm Clock Timers.*

(1) *Electric.*

(a) Assemble base, metal contacts, and alarm clock.

(b) Tie knot in one end of string. Thread other end through metal contacts and attach to alarm winding stem, which winches string and closes circuit.

Note. An alarm clock, being a very versatile delay, may be connected in many other ways.

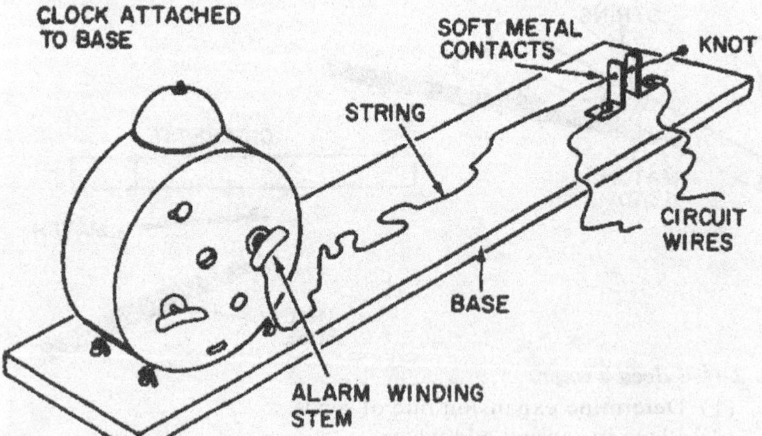

(2) *Nonelectric.*

(a) Drill hole in board of proper size to hold standard base tightly.
(b) Remove standard safety pin from firing device and replace with easily removed pin.
(c) Remove protective cap from standard base and crimp on nonelectric blasting cap.
(d) Screw standard base with blasting cap into firing device.
(e) Assemble alarm clock and firing device on board.
(f) Attach one end of length of string to eye in safety pin and the .other to alarm winding stem, which winches string and removes safety pin.

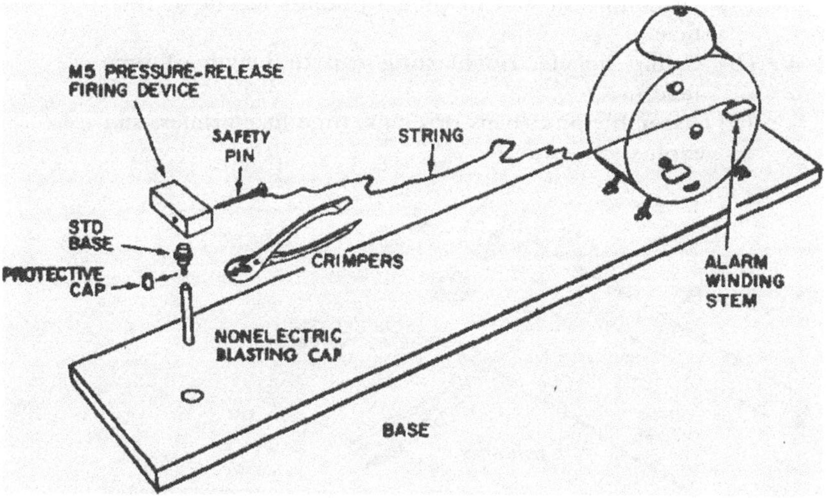

*d. **Wrist Watch Timer.***

(1) ***One-hour delay or less.***

(a) Drill small hole in plastic crystal and attach circuit wire with screw of proper length to contact minute hand.
(b) Attach other circuit wire to case.

(2) ***Twelve-hour delay or less.***

(a) Remove minute hand.
(b) Drill small hole in plastic crystal and attach circuit wire with screw of proper length to contact hour hand.
(c) Attach other circuit wire to case.

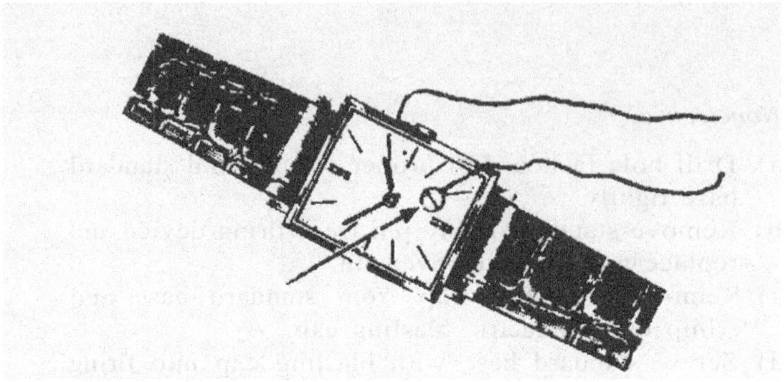

61. Bombs

a. Pipe Bombs.

(1) *Grenade.*

(a) Drill hole in cap or plug to admit length of time fuze.

(b) Crimp nonelectric blasting cap to length of time fuze.

(c) Assemble pipe, caps or plugs, time fuze primer, and explosive charge.

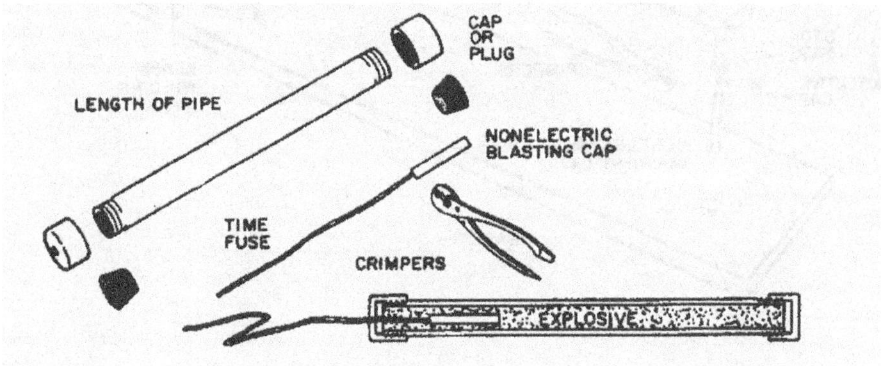

(2) *Antidisturbanee bomb.*

(a) Drill hole in end cap to admit length of burnt time fuze to make a bomb look like a "dud."

(b) Attach electric cap and mercury element on base.

(c) Test circuit with galvonometer *first,* then connect batteries with friction tape.

(d) Assemble bomb.

Caution: If possible, assemble bomb *in place,* as the mercury element, when disturbed, may cause premature explosion. To assemble more safely and easily, attach wrist watch timer in circuit.

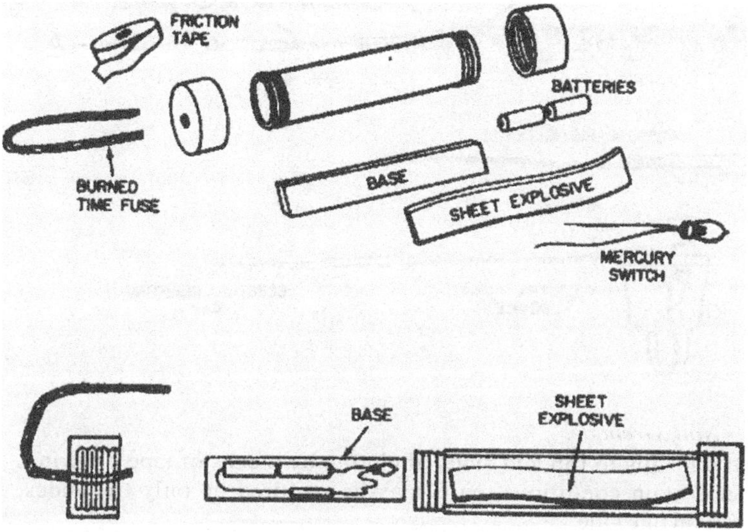

(3) *Shotgun bomb.*

(a) Close one end of pipe with hammer, allowing open- -ing for detonating cord primer or electric blasting cap.

(b) Remove protective cap from M1A1 pressure or M1 pull firing device and crimp on nonelectric blasting cap.

(c) Screw standard base with blasting cap into firing device.

(d) Assemble pipe, shrapnel, wadding, explosive, nonelectric primer or electric blasting cap (for controlled firing), and proper firing device.

Note. The force of the explosive and the strength of the pipe are important in calculating the size of the charge.

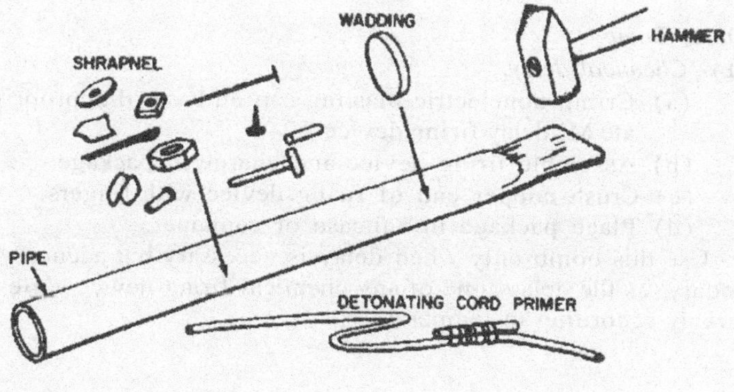

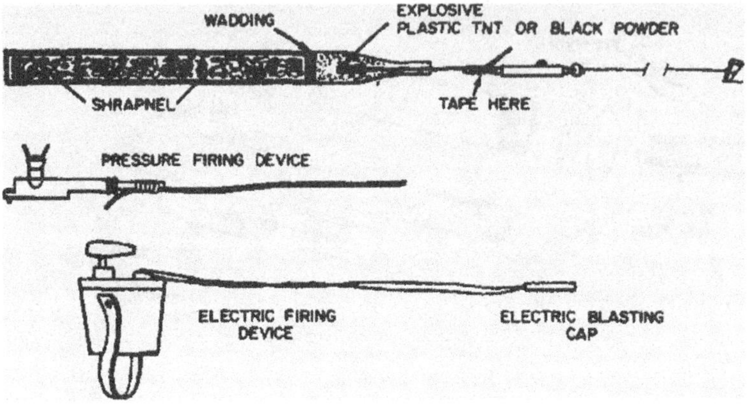

b. Nail Grenade.

Attach nails to top and sides of charge by means of tape or string. Under certain conditions, nails may be required on only two sides, or even on one side.

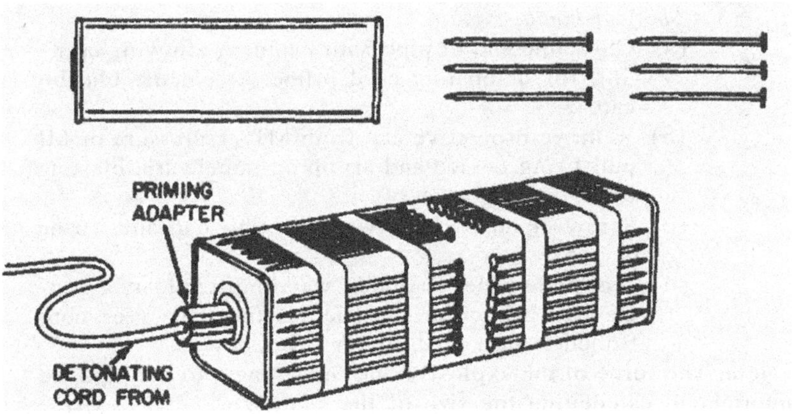

c. *Delay Bomb.*

(1) *Chemical delay.*

(a) Crimp nonelectric blasting cap on base of appropriate Ml delay firing device.

(b) Assemble firing device and charge in package.

(c) Crush copper end of firing device with fingers.

(d) Place package in suitcase or container.

Note. Use this bomb only when delay is necessary but accuracy is secondary, as the delay time of any chemical firing device varies considerably according to temperature.

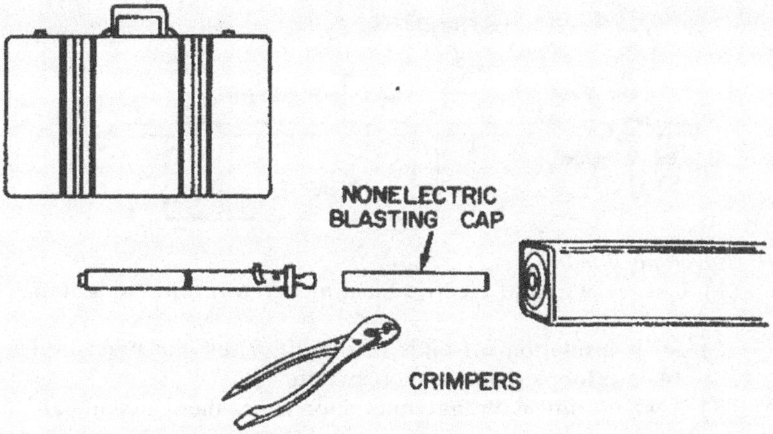

(2) *Alarm clock delay.*

(a) Drill hole in wooden base of proper size to hold standard base firmly.

(b) Remove standard safety pin from M5 pressure-release firing device and replace with easily-removed pin.

(c) Crimp nonelectric blasting cap on standard base and attach to firing device.

(d) Assemble alarm clock and firing device on wooden base.

(e) Attach one end of string in eye in pull pin and the other to the alarm winding stem so that its turning will winch the string and withdraw the pin.

(f) Place assembly in suitcase or container.

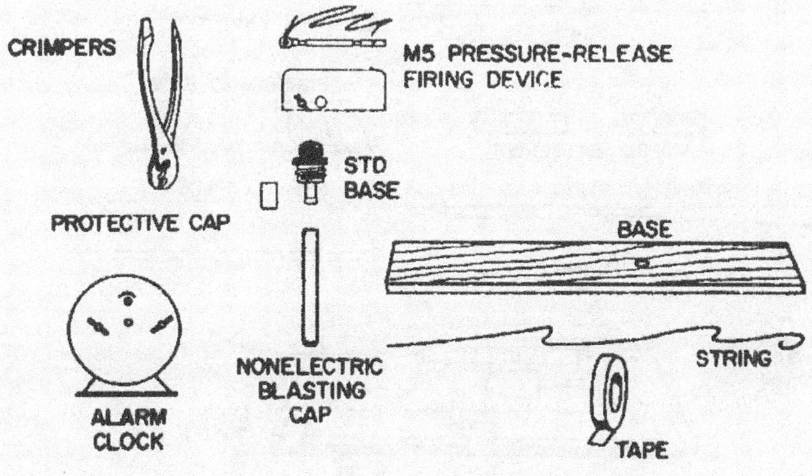

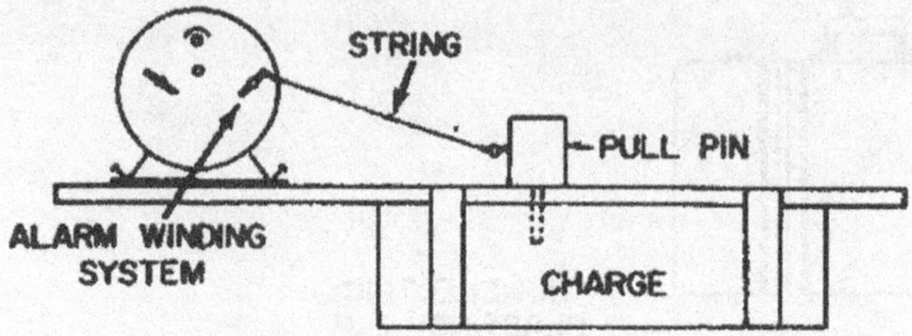

d. Envelope Bomb.

(1) Cut leg wires of electric blasting cap of proper length to make circuit.

(2) Strip insulation off ends of circuit wires and twist into 14,-inch loops to make loop switch.

(3) Test circuit with galvonometer *first*, then attach batteries.

(4) Assemble cardboard base, batteries, electric blasting cap, and explosive as package.

(5) Attach one end of string to loop switch so that it will pull the bared loops together to close circuit.

(6) Cut hole inside of envelope under flap.

(7) Fix package in envelope firmly and thread string through hole.

(8) Attach string firmly but concealed to underside of flap.

(9) Close envelope with elastic band.

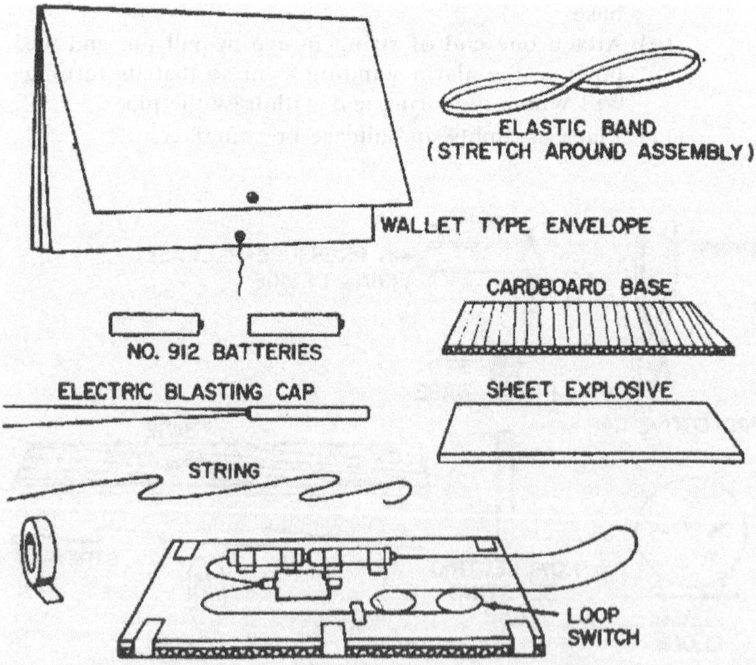

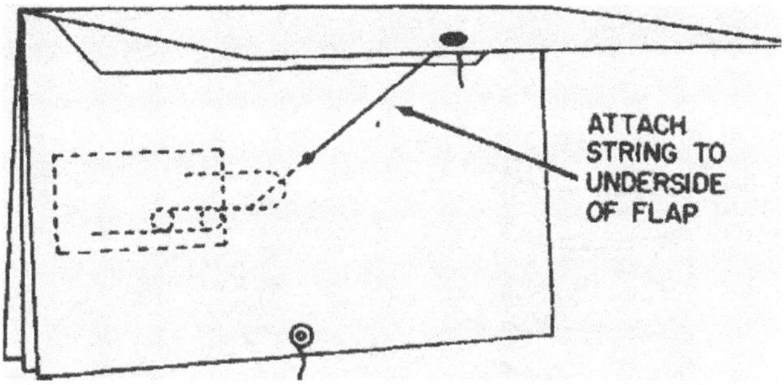

e. Hot Shrapnel Bomb.

(1) Remove protective cap from standard base and crimp on nonelectric blasting cap.

(2) Screw base with cap in Ml pull firing device.

(3) Crimp nonelectric blasting cap on one end of length of detonating cord, and install in Claymore mine.

(4) Attach firing device to detonating cord with tape.

(5) Assemble Claymore mine with priming and firing accessories and drum of napalm.

(6) Arm firing device.

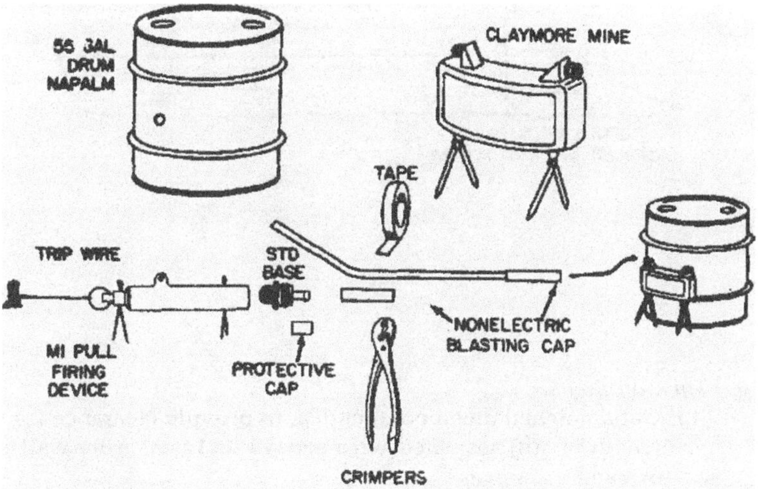

/. Rice Paddy Bomb.

(1) Remove protective cap from standard base and crimp on nonelectric blasting cap.

(2) Screw standard base with cap into Ml pull firing device.

(3) Assemble firing device, detonating cord, priming adapter, nonelectric blasting cap, and explosive charge.

(4) Attach charge to drum of napalm.

(5) Arm firing device.

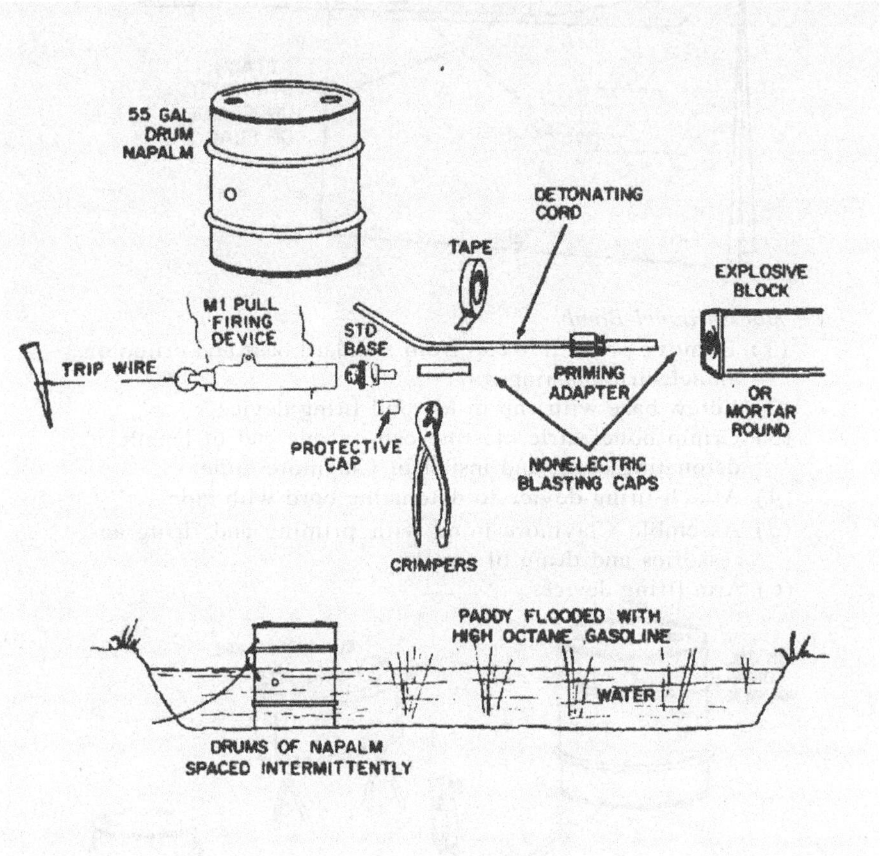

g. Tin Can Bomb.

(1) Cut a notched metal contact disk to provide clearance for length of stiff insulated wire and **Vi** to **14** in. from walls of can.

(2) Cut stiff insulated wire of proper length to support disk and strip insulation from both ends. Bend hook on one end to hold bare suspension wire.

(3) Bend stiff wire to proper shape.

(4) Assemble can, explosive, contact to can, blasting cap, insulated support wire, suspension wire and contact disk.

(5) Check circuit with galvonometer *first,* then connect batteries.

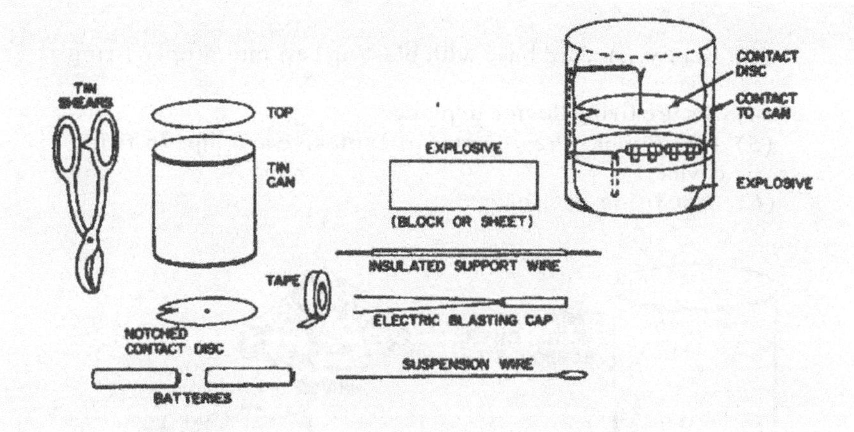

62. Miscellaneous Charges

a. *Improvised Shaped Charge.*

(1) Cut strip of thin metal to make cone of 30° to 60° angle to fit snugly into container.
(2) Place cone in container.
(3) Pack explosive firmly in container to a level of 2x height.
(4) Attach standoffs to set charge above target at height of of cone.
2x diameter of cone.
(5) Attach blasting cap at rear dead center of charge.

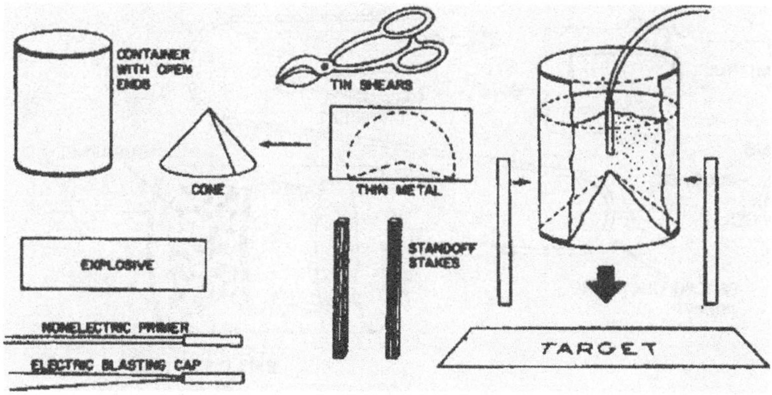

6. *Improvised Antipersonnel Mine.*

(1) Assemble container, explosive, separator, and shrapnel. *Explosive must be packed to uniform density and thickness* (should be Vi weight of shrapnel).
(2) Remove protective cap from standard base and crimp on nonelectric blasting cap.

(3) Screw standard base with blasting cap into proper firing device.
(4) Secure firing device in'place.
(5) Fix primer in rear center of explosive and tape to firing device.
(6) Arm firing device.

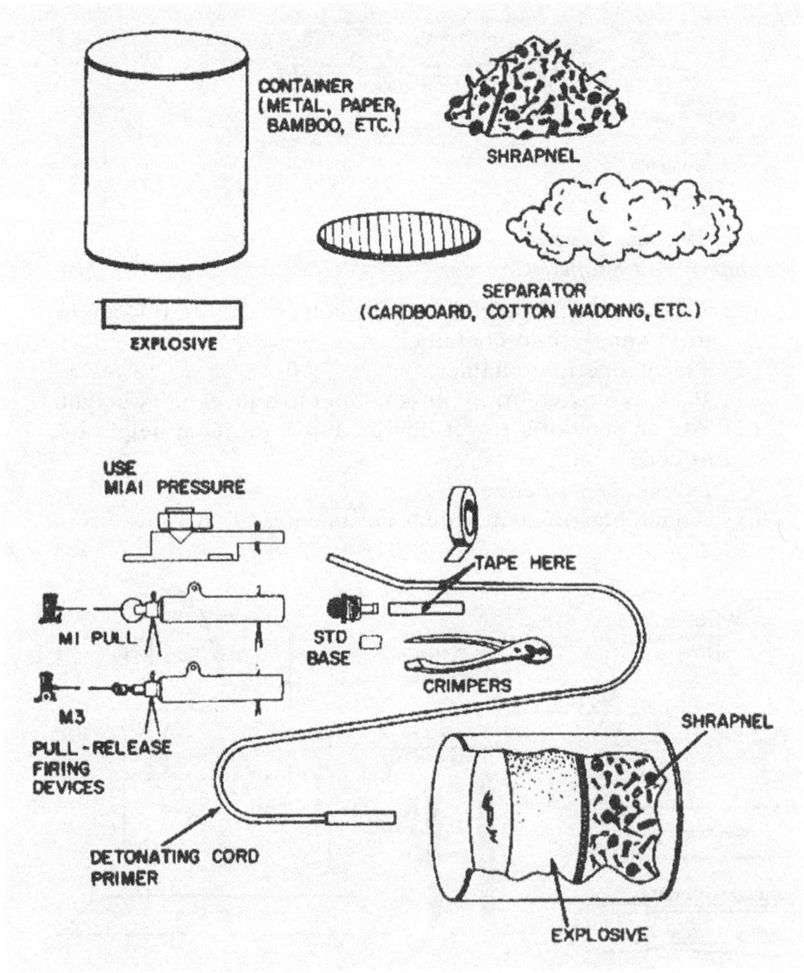

e. Platter Charge.

(1) Assemble container, charge, and platter. .Charge should weigh same as platter.
(2) Place primer in rear center of charge.
(3) Align center of platter with center of target mass.
(4) Attach and arm firing device.

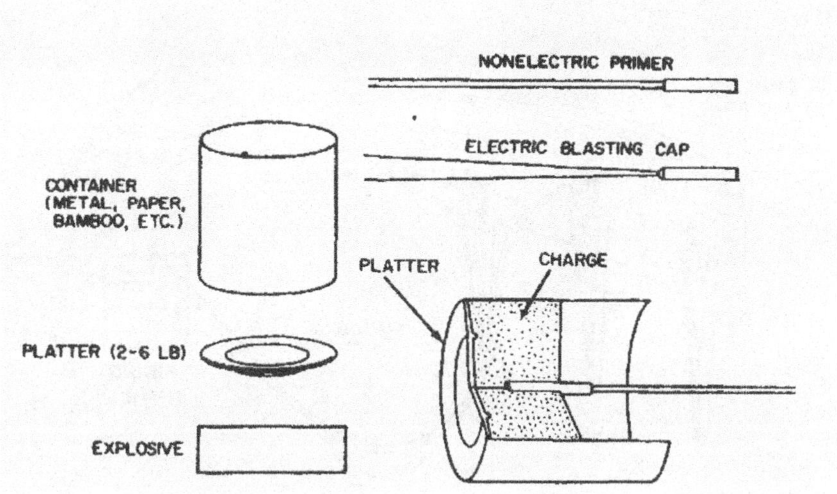

d. *Improvised Claymore.*

(1) Attach shrapnel to *convex* side of base and cover with cloth, tape, or screen retainer.
(2) Place layer of plastic explosive on *concave* side of base.
(3) Attach legs to *concave* side of base.
(4) Attach electric blasting cap at exact rear center.
(5) Attach firing device to firing wires at proper distance from mine for safety.

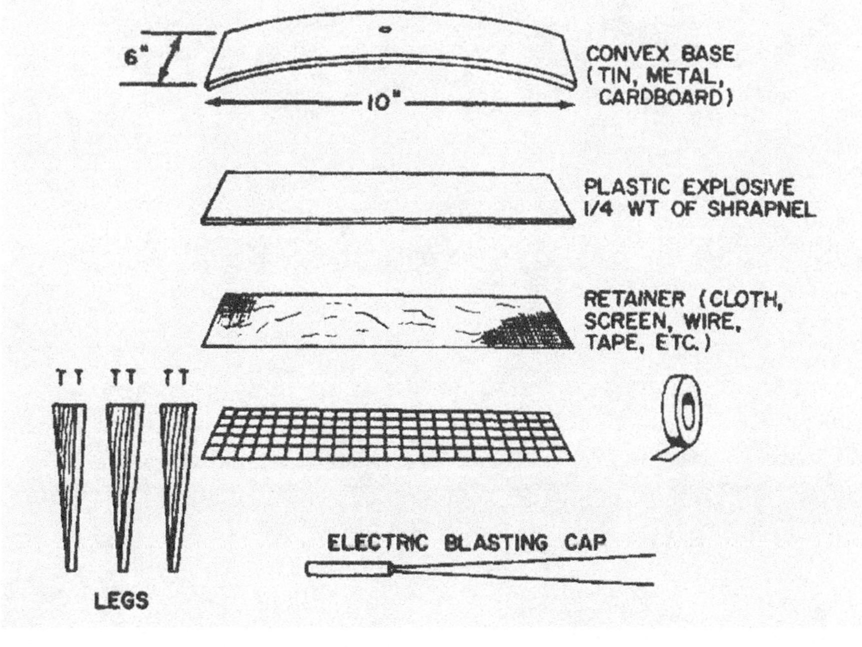

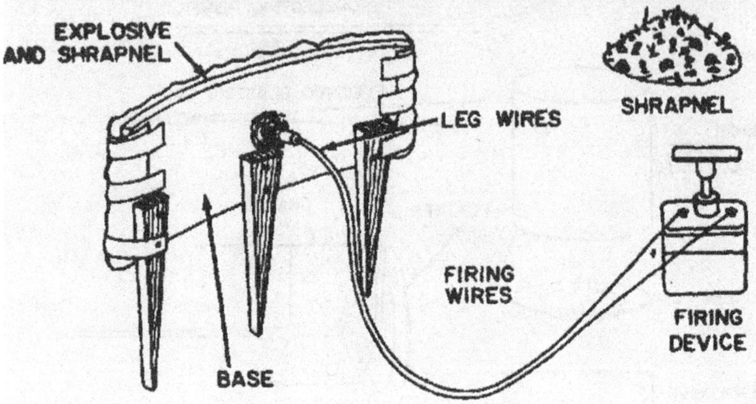
EXPLOSIVE
AND SHRAPNEL
LEG WIRES
SHRAPNEL
FIRING
WIRES
FIRING
DEVICE
BASE

CHAPTER 6

BOOBYTRAP DETECTION AND REMOVAL

Section I. CLEARING METHODS

63. Technicians

a. Although engineer and infantry specialists are responsible for boobytrap detection and removal, all military organizations assigned to combat zone missions must provide trained men to assist them.

b. If possible, trained engineer, infantry, or explosive ordnance disposal units will search out and neutralize all boobytraps in front of friendly troops or prepare safe passage lanes. When discovered, boobytraps will either be disarmed immediately or marked by warning signs. Only the simple ones will be disarmed during attack. Those more complicated will be marked and reported for removal.

c. To avoid casualty, boobytrapped areas, especially villages and other inhabited places, should be bypassed, to be cleared by specialists later. Tactical units will neutralize boobytraps only when necessary for continued movement or operation.

64. Clearance Teams

Men who clear boobytraps are organized into disposal teams and assigned to specific areas according to their training and experience.

a. Direction and control is the responsibility of the person in charge of clearance activities, who will-

(1) Maintain a control point near at hand and remain in close contact with his clearance parties.

(2) Give assistance to disposal teams when required.

(3) Preserve new types of enemy equipment found for more careful examination by engineer intelligence teams.

b. Searching parties will be sufficient in number to cover an area promptly, without interfering with each other.

c. In clearing a building, one person will direct all searching parties assigned.

d. Open area clearance will be preceded by reconnaissance if the presence of boobytraps is suspected. Once boobytraps are found, search must be thorough.

e. Searching parties must be rested frequently. A tired man, or one whose attention is attracted elsewhere, is a danger to himself and others working with him.

65. Tools and Equipment

a. *Body Armor.* Armor of various kinds is available. Special boots and shoe pacs, also issued, will give greater protection against blast than boots generally worn.

b. Mine Detectors.

(1) Three mine detectors useful in the removal of boobytraps are issued: AN/PRS-3 (Polly Smith) and the transistorized, aural indication model, designed for metal detection, and AN/PRS-4 for nonmetallic detection. Of the metal detectors, the transistorized model is the lighter and more powerful. All three models have the same deficiences. They may signal a small piece of scrap as well as a metal-cased explosive or signal an air pocket in the soil, a root, or disturbed soil generally.

(2) Operating time should not exceed 20 minutes to avoid operator fatigue. *Tired operators often become careless operators.*

c. Grapnels. These are hooks attached to a length of stout cord or wire, long enough for the operator *jo* pull a mine or boobytrap from place from a safe distance or from at least 50 meters behind cover.

d. Probes. Lengths of metal rod or stiff wire, or bayonets, are good probes for locating buried charges. Searching parties sometimes work with rolled-up sleeves better to feel trip wires and hidden objects.

e. Markers. Standard markers are carried by disposal teams to designate the location of known boobytraps, pending their removal.

f. Tape. Marking tape is useful for tracing safe routes and identifying dangerous areas.

g. Hand Tools. Small items, such as nails, cotter pins, pieces of wire, friction tape, safety pins, pliers, pocket knife, hand mirror, scissors, flashlight, and screw driver are very useful in boobytrap clearance.

66. Detection

a. The most careful observation is required for the detection of boobytraps. Soldiers must be trained and disciplined to be on guard, especially when moving over an area previously held by **the** enemy. Although a soldier may not be assigned the responsibility for their detection and clearance, he must be alert for any sign that **may** indicate their presence. He must also discipline himself to look carefully for concealed boobytraps before performing many **Acts** of normal life.

b. Often prisoners of war through interrogation give information on new or unknown boobytrap devices that may aid in their identification and handling later on. Local inhabitants also often provide information on boobytraps laid in the neighborhood.

e. Searching for boobytraps and delayed charges is difficult and tedious, particularly when intelligence is lacking or inadequate. The extent of search required, the **ease** of placing and camouflaging, and the great number of devices available to the enemy make the clearance of all charges almost impossible. Searching parties, before being sent out, will be briefed on all **that** is known about enemy activities in the area.

67. Outdoor Searching Techniques

As boobytraps are so deadly and as a rule cunningly conceived and hidden, outdoor searching parties should be suspicious of -

a. All moveable and apparently valuable and useful property.

b. All disturbed ground and litter from explosive containers.

e. Marks intentionally left behind to attract or divert attention.

d. Evidence of former camouflage.

e. Abrupt changes or breaks in the continuity of any object, such as unnatural appearances of fences, paint, vegetation, and dust.

f. Unnecessary things like nails, wire, or cord that may be part of a boobytrap.

g. Unusual marks that may be an enemy warning of danger.

h. All obstructions, for they are ideal spots for boobytraps. Search carefully before lifting a stone, moving a low hanging limb, or pushing aside a broken-down wheelbarrow.

i. Queer imprints or marks on a road, which may lead a curious person to danger.

j. Abandoned vehicles, dugouts, wells, machinery, bridges, gullies, denies, or abandoned stores. Also walk carefully in or around these as pressure-release devices are easily concealed under relatively small objects.

k. Areas in which boobytraps are not found immediately. Never assume without further investigation that entire areas are clear.

l. Obvious trip wires. The presence of one trip wire attached to an object does not mean that there are no others. Searching must be complete.

68. Indoor Searching Techniques

Those in charge of disposal teams should:

a. Assign no more than one man to a room in a building.

b. Indicate the finding of a large charge by a prearranged signal. All teams except those responsible for neutralizing large charges must then vacate the building immediately by the original route of entry.

c. Examine both sides of a door before touching a knob. Observe through a window or break open a panel. If doors and windows must be opened and both sides cannot be examined, use a long rope.

d. Move carefully in all buildings, for boobytraps may be rigged to loose boards, moveable bricks, carpets, raised boards or stair treads, window locks, or door knobs.

e. Never move furniture, pictures, or similar objects before checking them carefully for release devices or pull wires.

f. Never open any box, cupboard door, or drawer without careful checking. Sticky doors, drawers, or lids should be pulled with a long rope.

g. Not sit on any chair, sofa, or bed before careful examination.

h. Never connect broken wires or operate switches before checking the entire circuit. Such action may connect power to a charge.

t. Remove all switch plates and trace all wires that appear foreign to a circuit. Examine all appliances.

j. Investigate all repaired areas. Look for arming holes. Enlarge all wall and floor punctures. Cavities may be examined by reflecting a flashlight beam off a hand mirror. (This is also applicable for searching under antitank mines.)

k. Empty all fire boxes, remove the ashes, check fire wood, and move the coal pile.

l. Always work from the basement upward. Check, move, and mark everything movable including valves, taps, levers, controls, screens, and the like. A clockwork delay may not be heard if it is well hidden.

m. Double check basements and first floors—especially chimney flues, elevator and ventilator shafts, and insulated dead-air spaces. Check straight flues and shafts by observing from one end against a light held at the other. Dog-leg flues may be checked by lowering a brick from a safe distance.

n. Guard all buildings until they are occupied.

o. When possible and only after a thorough check, turn on all utilities from *outside* the building.

Note. A soldier by training can develop his sense of danger. Also by experience and careful continuous observation of his surroundings while in a combat area, he can develop an acute instinct that warns him of danger—a most valuable asset toward self-protection.

Section II. DISARMING METHODS

69. Neutralization

a. This is the making of a dangerous boobytrap safe to handle. If this is not possible, however, it must be destroyed. Neutralization involves two steps—*disarming* or replacing safeties in the firing assembly and *defuzing* or separating the firing assembly from the main charge and the detonator from the firing assembly.

b. Although types of boobytraps found in conventional warfare in a combat zone vary greatly, equipment used by most armies is basically similar except in construction details. Accordingly, a knowledge of the mechanical details and techniques in the use of standard U.S. boobytrapping equipment in conventional warfare prepares a soldier to some extent for dealing with that of the enemy. This, however is not true in guerrilla warfare. Most enemy boobytraps found recently in guerrilla infested areas, were cunningly and ingeniously improvised and laid. Such boobytraps can rarely be neutralized even by the most experienced specialists. These are discussed and illustrated in chapter 5.

c. Boobytraps may be neutralized by two methods. (1) Whenever the location permits, they may be destroyed by actuating the mechanism from a safe distance or detonating a charge near the main charge. These should be used at all times unless tactical conditions are unfavorable (2) When necessary, boobytraps may be disassembled by hand. As this is extremely dangerous, it should be undertaken only by experienced and extremely skillful specialists.

Note. Complete knowledge of the design of the boobytrap should be obtained before any neutralization is attempted.

d. In forward movements, all complicated mechanisms found are bypassed. These are marked and reported for neutralization later, when more deliberate action may be taken without harrassment by enemy fire.

e. All boobytraps exposed to blast from artillery fire or aerial bombing should be destroyed in place.

f. Boobytraps with unrecognizable or complicated firing arrangements should be marked and left for specialists to disarm.

(1) Electrically fired boobytraps are among the most dangerous **of** all. Though rare in the past, they now turn up frequently in guerrilla warfare. Some may be identified by the presence of electric lead wires, dry cells, or other batteries. Some are small containers with all elements placed inside which actuate at the slightest disturbance. These can hardly be disarmed even by experts.

(2) Another difficult type has delay fuzing—a spring-wound **or** electric clockwork for long delay periods or chemical action firing devices. As the time of detonation is uncertain, such boobytraps should be destroyed in place, if possible or tactically fesasible.

70. Rules of Conduct

a. Keep in constant practice by inspecting and studying all known boobytrap methods and mechanisms.

b. Develop patience. A careless act may destroy you and others as well.

e. Remember that knowledge inspires confidence.

d. Let only one man deal with a boobytrap. Keep all others out **of** danger.

e. If in doubt, get help from an expert.

f. Never group together when there is danger.

g. Be suspicious of every unusual object.

h. Regardless of nationality, consider every enemy a ruthless, cunning and ingenious killer.

71. Detailed Operations

a. *Destruction in Place.*

(1) If damage is acceptable, which is generally the case out **of** doors, the operator may initiate boobytrap riggings by their own mechanism or by a rope from a safe position (at least 50 meters away).

(2) The easiest method of getting rid of a boobytrap is to detonate a pound of high explosive adjacent to the main charge.

b. Removal of Main Charge (Antitank Mine).

Careful probing or search around the charge is necessary **to** locate and neutralize all antilift devices. Recognition of **the type of** firing mechanisms used is necessary to avoid casualty. **All safety** devices must be replaced. If complete neutralization seems doubtful, the charge should be pulled from place by a grapnel or rope from a safe location. After the charge is pulled, the operator should wait at least 30 seconds as a safeguard against a concealed delay action fuze.

c. Hand Disarming. None but trained specialists should undertake this job, unless the boobytrap's characteristics and disarming techniques' are well known. Trained specialists only should inspect and destroy all unusual or complicated mechanisms for safety reasons and for information on new enemy devices. The following procedures for hand neutralization are for guidance only, as the exact sequence depends on the type of device and the manner of placement.

(1) Do not touch any part of a boobytrap before examining it thoroughly. Locate all firing devices and their triggering mechanisms.

(2) When tracing wires, look for concealed intermediate devices laid to impede searching. Do not disturb any wires during the examination of the boobytrap.

(3) Cut loose trip wires only after careful examination of all connecting objects and their functions and replacing all safeties.

(4) Trace taut wires and disarm all connected firing devices by replacing safeties. Taut wires should be cut only when the danger at both ends has been eliminated.

(5)) Replace safeties in all mechanisms, using nails, lengths of of wire, cotter pins, and other objects.

(6) Never use force in disarming firing devices.

(7) Without disturbing the main charge, cut detonating cord or other leads between the disarmed firing devices and the main charge.

(8) Cut wires leading to an electric detonator—*one at a time.*

(9) When using a probe, push it gently into the ground. *Stop* when you touch any object. It may be a pressure cap or plate.

(10) Once separated, boobytrap components should be removed to a safe storage or disposal area.

d. Special Precautions.

(1) Be very cautious in handling delay mechanisms. Although there may be little danger before the appointed time, auxiliary firing devices may be present. All complicated and confusing deyices should be destroyed in place or marked for treatment by specialists.

(2) Explosive containers of wood or cardboard, buried for long periods are dangerous to disturb. They are also extremely dangerous to probe if in an advanced state of decomposition. Deteriorated high explosives are very susceptible to detonation. Thus destruction in place of a boobytrap and in a concentrated area long exposed to moisture may detonate many others simultaneously.

(3) Metallic explosive containers, after prolonged burial, are often dangerous to remove. Oxidation may make them resistant to detection. After a time the explosive may become contaminated, increasing the danger in handling. Explosives containing picric acid are particularly dangerous as deterioration from contact with metal forms extremely sensitive salts readily detonated by handling.

(4) Fuzes of certain types become extremely sensitive to disturbance from exposure to wet soil. The only safe method of neutralizing or removing such deteriorated boobytraps is detonation in place.

72. Explosive Disposal

a. Usually, explosive items recovered by hand neutralization are destroyed by specially-trained explosive ordnance disposal units. Should untrained troops be required to do this, they should follow established procedures with great care. Explosives to be detonated should be buried in a pit at least 4 feet deep under 2 feet of earth, free of rocks or other matter that may become flying debris.

b. Components should be placed on their side or in position to expose their largest area to the force of the initiating explosive. Demolition blocks should be used for destruction of these components, if available; but bangalore torpedoes or dynamite may be substituted. Primed charges should always be connected to firing mechanisms by detonating cord, so that blasting caps may be connected at the last minute. This eliminates opening the pit in the event of a misfire. All persons engaged in disposal should take cover when explosive components are detonated. Despite the 2-foot layer of earth, fragments may be thrown at high velocity for several hundred yards.

INDEX

TECHNICAL MANUAL
NO. 9-1005-224-10

***TM 9-1005-224-10**
T.O. 11W2-6-4-11
HEADQUARTERS
DEPARTMENTS OF THE ARMY
AND AIR FORCE
Washington D.C., 2 April 1998

Operator's Manual for MACHINE GUN, 7.62-MM, M60 W/E

REPORTING ERRORS AND RECOMMENDING IMPROVEMENTS

You can help improve this manual. If you find any mistakes or if you know of a way to Improve the procedures, please let us know. Mail your letter or DA Form 2026 (Recommended Changes to Equipment Publications and Blank Forms) direct to: Director, Armament and Chemical Acquisition and Logistics Activity, ATTN: AMSTA-AC-NML, Rock Island, IL 61299-7630. A reply will be furnished to you.

*This manual supersedes TM 9-1005-224-10, dated 30 July 1985; TM 9-1005-224-10HR, dated 18 May 1979; TB 9-1005-224-50-1, including all changes.

WARNING SUMMARY

Be sure to clear weapon before starting inspection.

If cover is opened on a hot cartridge (hot barrel), an open cover cookoff could occur and result in serious injury or death. Evacuate area for 15 minutes and then do REMEDIAL ACTION. Refer to WP 00011 00.

Never reload a runaway machine gun until it has been repaired. Be sure machine gun is cleared before moving it.

When removing a stuck unfired cartridge, stay clear of the muzzle. Do not allow cartridge to contact any hard surface. Cartridge may fire on contact. Remove a stuck unfired cartridge using the same procedures for removing a stuck or ruptured cartridge case.

Before field stripping check that the bolt is in forward position.

To prevent possible bodily injury, or aircraft/weapon damage, personnel should not stand the weapon on its barrel assembly when disassembling or assembling the weapon.

WARNING SUMMARY (cont)

Bolt assembly in under spring tension; it can twist and injure your hand.

Barrels issued for a specific gun will not be changed from gun to gun. Each barrel and bolt assembly should be tagged during cleaning with the serial number of the receiver (gun) and should remain together as initially assigned. Interchanging barrels/bolts may result in injury or death.

The climate temperature in different regions will make a difference as to what constitutes a hot gun. A hot, sunny day can cause a cookoff within 50 rounds, weapon and ammunition in the sun.

Insert the hinge pin from one side and the hinge pin latch from the opposite side.

After unloading dummy round, be sure the barrel is clear.

Ammunition to be used in machine gun is shown in WP 0031 00.

Keep weapon pointed at target - down range/impact area.

WARNING SUMMARY (cont)

Keep safety on 'S' (safe) until your ready to fire.

Always check the chamber/bore after clearing the weapon

Check barrel bore and chamber before firing.

Do not interchange bolt/barrel assemblies from one weapon to another. This may cause injury or death to personnel.

Change hot barrels with your heat protective mitten only or by holding bipod legs. (Refer to WP 0017 00 for additional hot barrel procedures.)

Do not interchange flat leaf springs from one model weapon to another.

Check that assigned and spare barrels have been headspaced and tagged to your receiver. Rotate usage of the barrel.

WARNING SUMMARY (cont)

Do not fire blank ammunition toward personnel within 20 feet of the muzzle. Fragments of a closure wad or particles of unburned propellant might inflict injury within that range.

Dry cleaning solvent is FLAMMABLE and TOXIC and must be kept away from open flames and used in a well ventilated area. Use of rubber gloves is necessary to protect the skin when washing machine gun parts.

Appropriate eye protection is recommended when cleaning your weapon and/or its parts.

For first aid information see FM 21-11.

TABLE OF CONTENTS

CHAPTER 1

INTRODUCTION

GENERAL INFORMATION 0001 00

SCOPE

Type of Manual: Operator's Manual.

Model Numbers and Equipment Names: M60, 7.62mm, Machine Gun, and M60D, 7.62mm, Machine Gun.

Purpose of Equipment:

The M60, 7.62mm, Machine Gun is a general purpose weapon capable of being fired from several mounts or handheld. The weapons primary use is for ground operations.

The M60D, 7.62mm, Machine Gun is a general purpose weapon capable of being fired from several mounts. The weapons primary use is for support of ground operations. The M60D is an aircraft door-mounted or vehicle-mounted machine gun.

MAINTENANCE FORMS AND RECORDS

Department of the Army forms and procedures used for equipment maintenance will be those prescribed by DA PAM 738-750, the Army Maintenance Management System (TAMMS).

REPORTING EQUIPMENT IMPROVEMENT RECOMMENDATIONS (EIR's).

If your machine needs improvement, let us know. Send us an EIR. You, the user, are the only one who can tell us what you don't like about your equipment. Let us know why you don't like the design or performance. Put it on an SF 368 (Quality Deficiency Report). Mail it to us at Commander, US Army Armament Research, Development and Engineering Center, ATTN: AMSTA-AR-QAW-A, Rock Island, IL 61299-7630. We'll send you a reply.

EQUIPMENT DESCRIPTION AND DATA 0002 00

EQUIPMENT CHARACTERISTICS, CAPABILITIES AND FEATURES.

The M60 and M60D are air-cooled and have fixed headspace allowing quick barrel changes for cooling and maintenance when required. In order to extend the life of the barrels, retain accuracy, and allow for continuous firing over long periods of time, two barrel assemblies are issued with each gun. The machine guns are gas-operated and fire from open bolt position.

LOCATION AND DESCRIPTION OF MAJOR COMPONENTS.

M60 and M60D Machine Gun.

A **COVER ASSEMBLY** - Positions and holds cartridge in place for stripping, feeding link belt, and chambering rounds.

B **COCKING HANDLE ASSEMBLY** - Provides a means to manually move the bolt assembly to the rear.

C **TRIGGER MECHANISM AND GRIP ASSEMBLY (M60 ONLY)** - Controls the firing of the machine gun.

D **SEAR AND SAFETY HOUSING (M60D ONLY)** - Controls the firing of the machine gun.

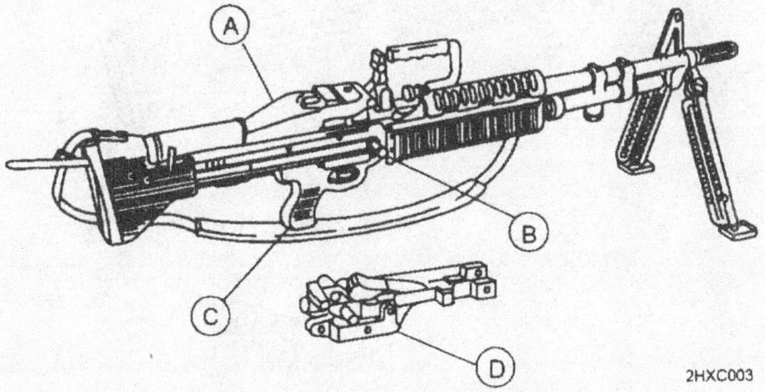

LOCATION AND DESCRIPTION OF MAJOR COMPONENTS (cont)

E **BARREL AND BIPOD ASSEMBLY -** The barrel houses cartridges for firing. The bipod assembly provides a semistable platform when the machine gun is fired from the prone position.

F **CARRYING HANDLE ASSEMBLY** - Provides a means to carry the machine gun with one hand. The carrying handle assembly folds down when the rear sight is used and the machine gun is fired.

G **RECEIVER ASSEMBLY** - Supports all major components. Houses internal parts and, through a series of cam ways, controls operation of weapon.

H **SHOULDER GUN STOCK (M60 ONLY)** - Provides a suitable surface to stabilize the weapon against the shoulder while firing the machine gun from any position except from the hip.

I **GRIP AND TRIGGER ASSEMBLY (M60D ONLY)** - Provides handles to move machine gun and houses the machine gun trigger.

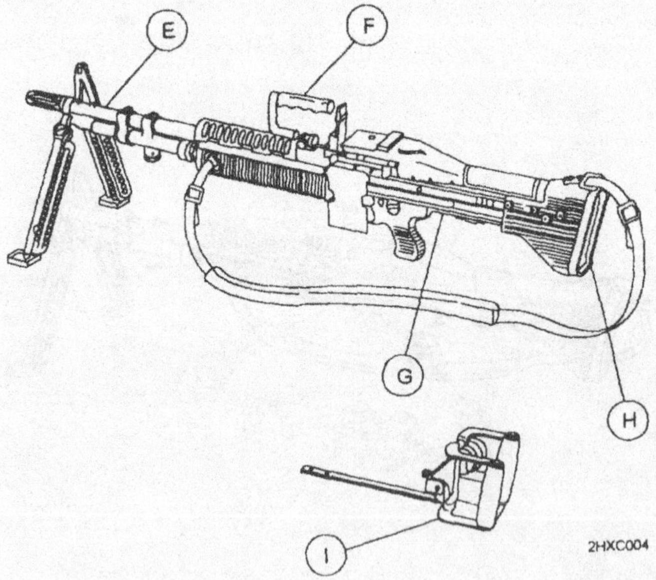

LOCATION AND DESCRIPTION OF MAJOR COMPONENTS (cont)

J **CARTRIDGE FEED TRAY AND HANGER ASSEMBLY (M60 ONLY)** - Guides cartridges for positioning and feeding. The hanger assembly supports the bandoleer.

K **CARTRIDGE TRAY ASSEMBLY (M60D ONLY)** - Guides cartridge for positioning and feeding.

L **REAR SIGHT ASSEMBLY (M60 ONLY)** - Provides a means to aim the machine gun at the target with accuracy. The sight adjusts horizontally as well as vertically.

M **REAR SIGHT ASSEMBLY (M60D ONLY)** - Provides a means to aim the machine gun in the general area of the target. The rear sight is nonadjustable.

N **FOREARM ASSEMBLY (M60 ONLY)** - Provides a hand hold when firing from the hip or from a standing or kneeling position.

O **SMALL ARMS SLING (M60 ONLY)** - Used for support during assault firing and for transport.

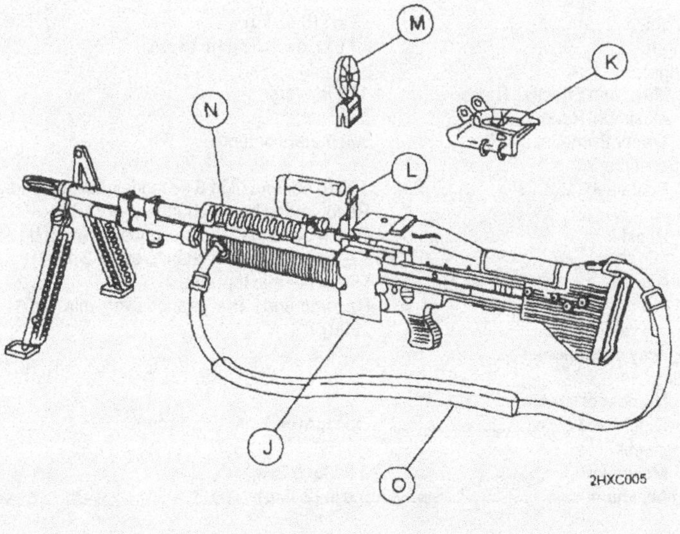

DIFFERENCES BETWEEN MODELS - Differences between the M60 and M60D machine gun are listed below.

Assembly/Component	M60	M60D
Tray and hanger assembly	X	
Cartridge feed tray assembly		X
Dust and moisture seal boot		X
Forearm assembly	X	
Grip and trigger assembly		X
Gun adapter		X
Magazine hanger assembly		X
Rear sight (adjustable)	X	
Rear sight (nonadjustable)		X
Receiver assembly	X	
Receiver assembly		X
Sear and safety housing assembly		X
Sear assembly link and spring		X
Shoulder gun stock	X	
Small arms sling	X	
Trigger mechanism grip assembly	X	
Quick release pin		X

EQUIPMENT DATA

M60 Machine Gun.

Weight.. 23 lb (10.43 kg)
Length.. 43.5 in. overall (1.1m overall)
Range:
 Maximum Effective Range 1100meters
 Maximum Range .... 3725 meters
 Tracer Burn Out .... 900 meters or more
Rates of fire:
 Sustained .... 100 rd per min (4 to 6 seconds between burst)
 Recommend barrel change every 10 min
 Rapid .... 200 rd per min (2 to 3 seconds between burst)
 Recommend barrel change every 2 min
 Cyclic .... 550 rd per min (approx)
 Recommend barrel change every min
Muzzle velocity .... 2800 FPS
Capacity of bandoleer .... 100 rds
Rifling:
 Number of lands.......................... 4
 Right hand twist.......................... One turn in 12 in (30.54 cm)
Trigger pull:
 Maximum.................................. 11.5 lb (5.2 Kg)
 Minimum.....................................6.0 lb (2.7 kg)

EQUIPMENT DATA (cont)

M60D Machine Gun

Weight	25 lb (11.33 kg)
Length	43.5 in overall (1.1m overall)
Range:	
Maximum Effective Range	1100 meters
Maximum Range	3725 meters
Tracer Burn Out	900 meters or more
Rates of fire:	
Sustained	100 rd per min (4 to 6 seconds between burst) Recommend barrel change every 10 min
Rapid	200 rd per min (2 to 3 seconds between burst) Recommend barrel change every 2 min
Cyclic	550 rd per min (approx) Recommend barrel change every min
Rifling:	
Number of lands	4
Right hand twist	One turn in 12 in (30.54 cm)
Trigger pull at sear activator:	
Maximum	20 lb (9.06 kg)
Minimum	10.5 lb (4.75 kg)

CHAPTER 2

OPERATING INSTRUCTIONS

This page intentionally left blank.

DESCRIPTION AND USE OF OPERATOR'S CONTROLS AND INDICATORS 0003 00

M60 MACHINE GUN

COCKING HANDLE. Pull back to move bolt rearward.

COVER LATCH. Holds cover closed. Pull back on lower end of latch to open.

BARREL LOCK LEVER. Locks barrel in place. To unlock barrel, push in on barrel lock lever from left side of weapon and raise lever.

BIPOD LEG PLUNGER. Depress to extend or retract bipod foot.

TRIGGER. Pull rearward to fire weapon. Release to stop firing.

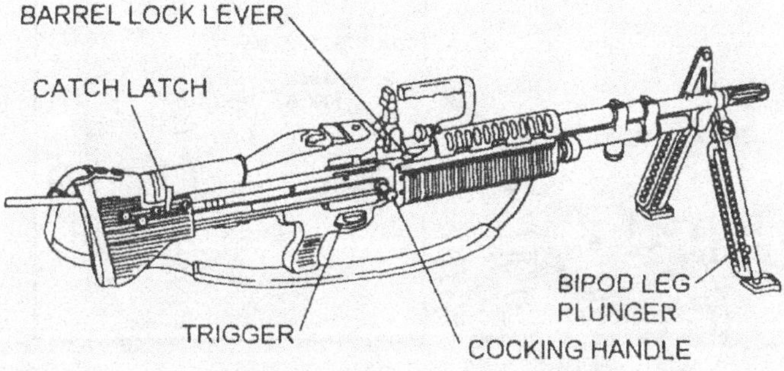

M60

M60 MACHINE GUN (cont)

SAFETY. Move safety up to 'F' (fire) and down to 'S' (safe).

REAR SIGHT. Adjustable to 1100 meters.

- Range Scale - graduated from 300 to 1100 meters.
- Scale Retaining/Adjusting Screw - locks range scale in place.
- Elevating Knob - use for minor (fine) adjustments in range.
- Windage Scale - adjust for left or right wind.
- Rear Sight Aperture - lines up with front sight for long range targets.

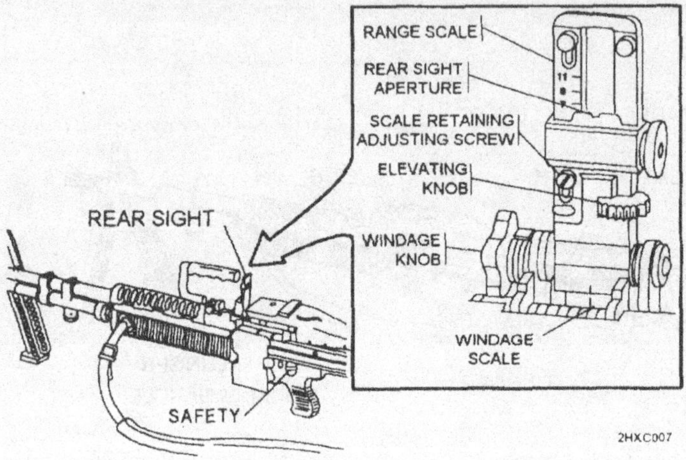

M60D MACHINE GUN

GRIP AND TRIGGER ASSEMBLY. Pull back on both triggers to fire. Release to stop firing.

SAFETY. Push to right for safe 'S', to left for fire 'F'.

REAR AND FRONT SIGHTS. Aim through the ring sight (rear sight) and front sight to line up target. Fire. Adjust range by watching tracers.

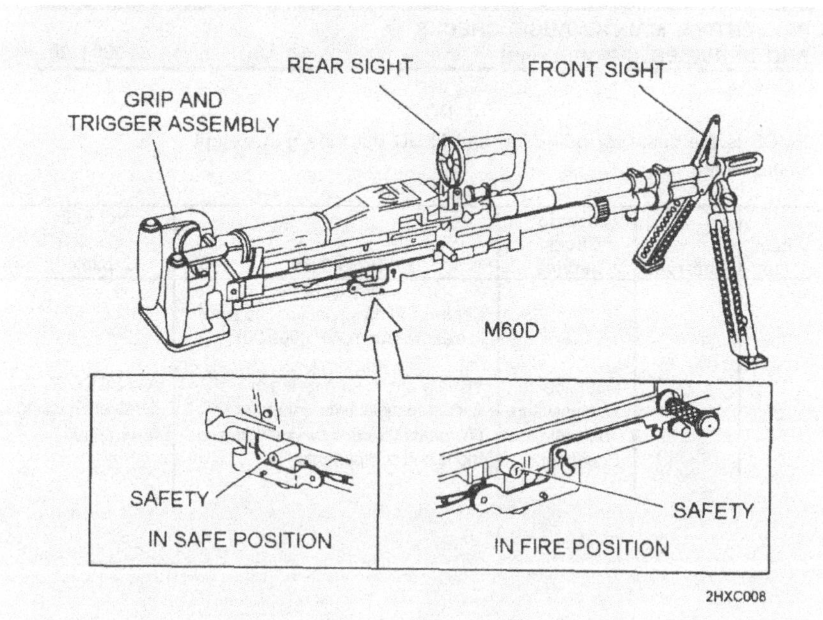

PREVENTIVE MAINTENANCE CHECKS AND SERVICES (PMCS) 0004 00

GENERAL

Always keep in mind the CAUTIONS and WARNINGS when performing your before (B) or during (D) PMCS. The numbers in the item number column shall be used for the "TM Number" column on DA Form 2404, Equipment Inspection and Maintenance Worksheet, in recording results of PMCS. After you operate, perform your after (A) PMCS.

Perform BEFORE PMCS if: (1) you are the assigned operator and the machine gun has been stored and not used for a period of 90 days or(2) you have been issued the machine gun for the first time.

WARNING

Do not interchange barrel assembly or bolt assembly from one machine gun to another. Doing so may result in injury or death.

The climate temperature in different regions will make a difference as to what constitutes a hot gun. A hot, sunny day can cause a cookoff within 50 rounds, weapon and ammunition in sun.

NOTE

PMCS is the same for both M60 and M60D machine gun except where noted.

Item No.	Interval	Item to Check/ Service	Procedure	Not Fully Mission Capable if:
			WARNING Before starting inspection, be sure to clear weapon (WP 0009 00).	
1	B	M60/M60D Machine Gun Tools and Equipment	Visually check for missing or damaged tools and equipment (WP 0034 00/0035 00) including tripod and components.	Any BII/COEI items are missing ordamaged.

Item No.	Interval	Item to Check/ Service	Procedure	Not Fully Mission Capable If:
2	B	Barrel Assemblies	Open cover, pull bolt to rear, push cocking handle forward and lock. Close cover. place on safe (S). Tilt weapon end to end to make sure gas piston slides freely (a clicking sound will be heard as the piston slides In the cylinder). Run a dry swab through the bore to remove excess lubricant or obstructions.	Gas piston will notslide. Obstruction in barrel. Gas cylinder vent, cylinder plug, and cylinder nut loose.
3	B	Trigger Mechanism Grip Assembly (M60)/ Sear and Safety Housing Assembly (M60D)	Make sure flat leaf spring is in place (WP 0018 00). Check that safety is in 'F' position. Pull bolt to rear, move safety to 'S'. Push cocking handle forward and lock and pull the trigger.	Weapon functions with safety at 'S'.

Item No.	Interval	Item to Check/ Service	Procedure	Not Fully Mission Capable If:
4	B	Cover Assembly	Lightly lubricate all moving parts with lubricant on a clean swab (item 6, WP 0038 00).	
5	B	Gun Receiver Assembly	Check for free movement of cocking handle. Check barrel lock lever. Make sure it locks in place.	Cocking handle binds or barrel is loose.
6	D	M60/M60D Machine Gun	Erratic or sluggish firing may indicate carbon buildup. Switch barrels (WP 0018 00). If firing is still erratic or sluggish, field-strip the weapon and perform the necessary maintenance to complete the tire mission.	Weapon ceases to operate.

Item No.	Interval	Item to Check/ Service	Procedure	Not Fully Mission Capable If:
7	D	Small Arms Sling (M60)	Inspect sling for damage and for missing or loose sling swivels.	Swivels loose or missing. Sling is damaged.
8	A	M60/M60D Machine Gun	Fieldstrip, clean, inspect and lubricate entire weapon (WP 0018 00).	
9	A	M60/M60D Machine Gun (Inspection)	Inspect operating rod assembly and barrel assembly for burrs, cracks and chips. Make sure the gas piston slides back and forth freely. Inspect components of trigger assembly for wear. Place safety in 'F' position, sear should move slightly. Inspect breechbolt assembly for defects (e.g., cracks or chips). Reassemble weapon and check for proper functioning.	Weapon fails to function or has damaged or missing parts.

OPERATION UNDER USUAL CONDITIONS - LOADING 0005 00

NOTE

The following procedures illustrate the M60 machine gun except where the M60 parts differ from the M60D park, then both weapons are illustrated.

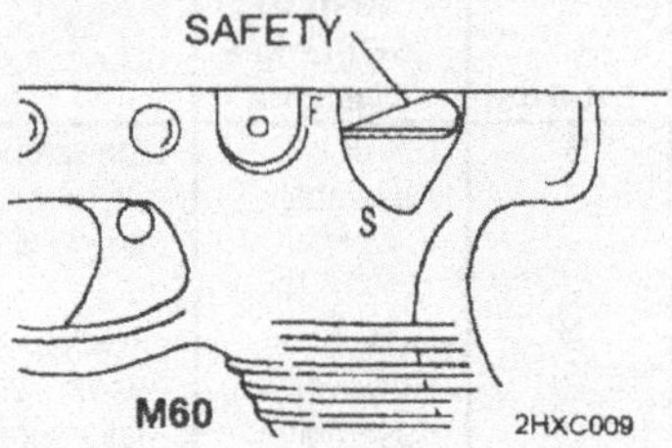

1 Point weapon down range.

2 Safety should be on F.
- M60 - Move safety to F.
- M60D - Push in on F.

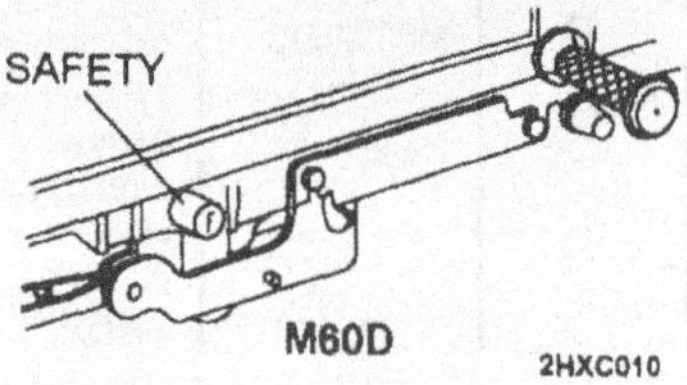

OPERATION UNDER USUAL CONDITIONS - LOADING (cont) 0005 00

3 Pull cocking handle rearward, then push it forward until it locks.

4 M60 - Move safety to S.
M60 D - Press in on S.

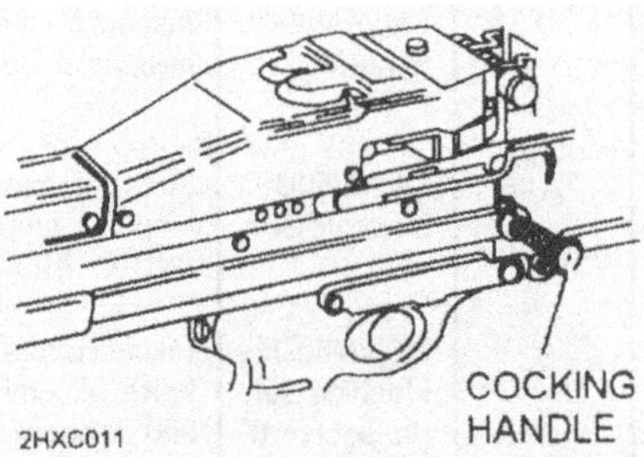

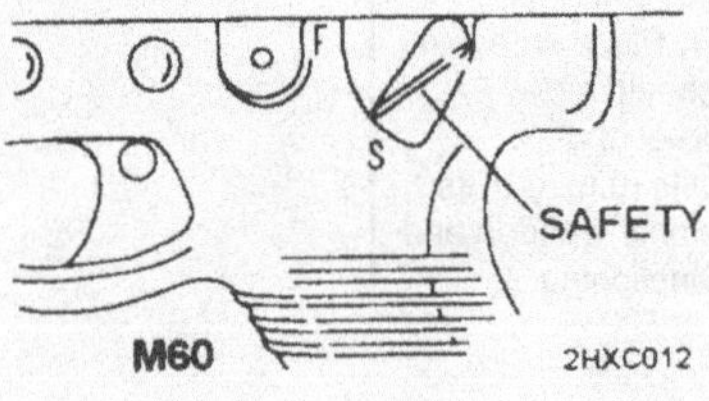

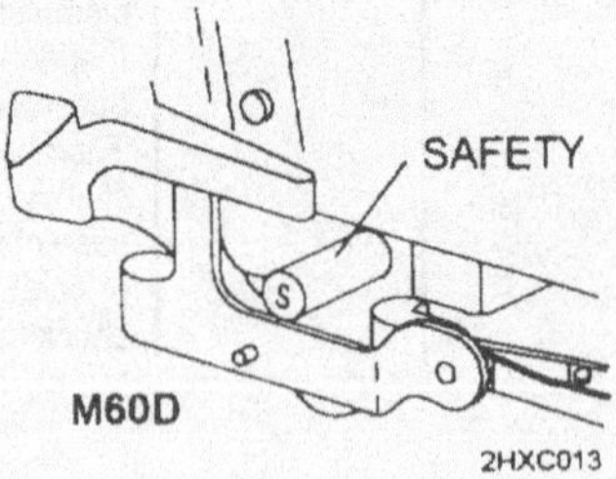

CAUTION

Bolt assembly must be rearward when opening or closing the cover.

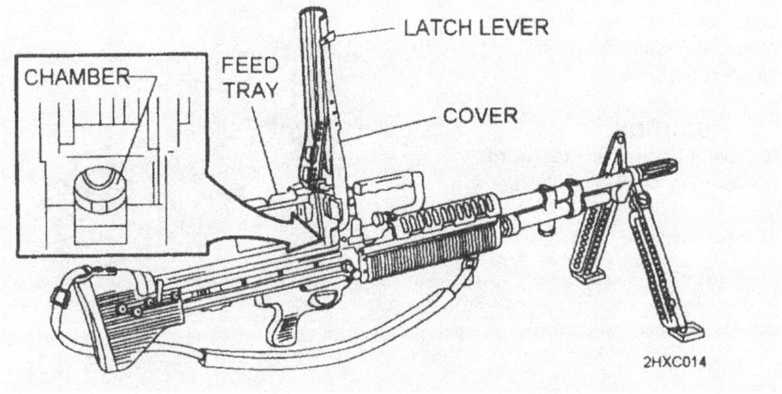

5 Turn latch lever and open cover.

6 Raise cartridge feed tray and check chamber for ammunition.

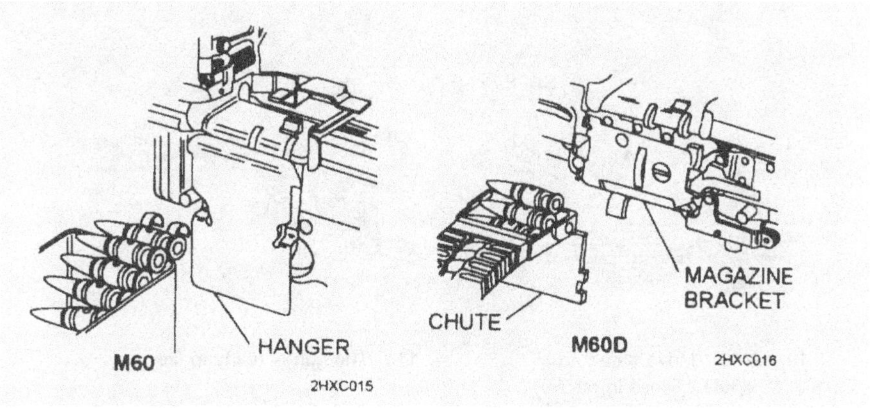

7 M60 - Place bandoleer on bandoleer hanger.

M60D - Attach ammunition chute to magazine bracket.

8 Place ammunition on feed tray, with open side of links down and with first round in the feed groove. (Make sure cartridge retainer pawl holds ammunition link on feed tray.)

CAUTION

Bolt assembly must be rearward when opening or closing the cover.

9 Close cover.

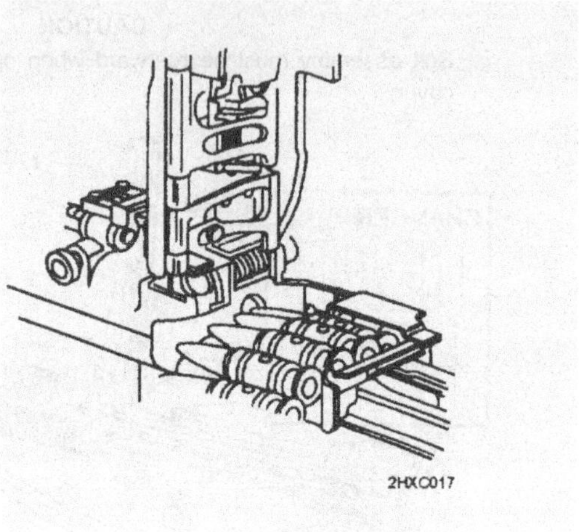

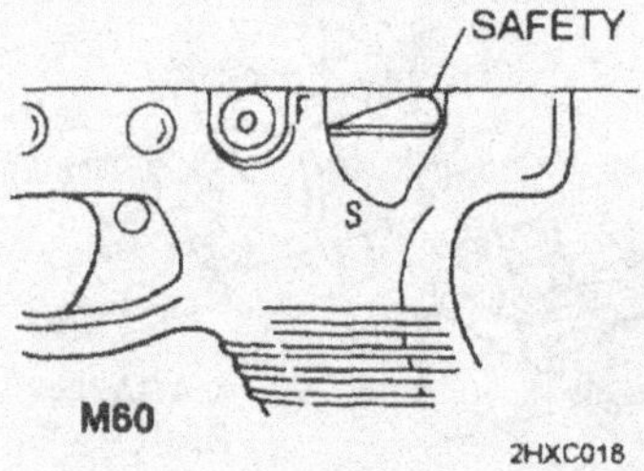

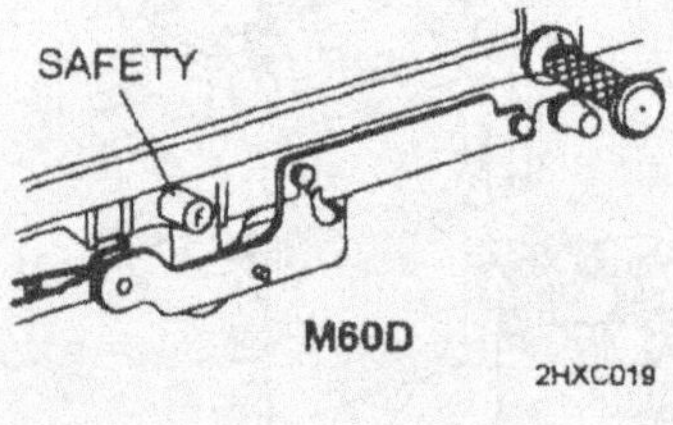

10 M60 - Move safety to 'F'.
M60D - Press in on 'F'.

11 The gun is ready to fire.

OPERATION UNDER USUAL CONDITIONS - HOW TO PUT THE M60 ON THE M122 TRIPOD 0006 00

1 Unfold front leg and spread rear legs until leg lock engages.

2 Insert pintle assembly. Rotate pintle lock release cam to lock (see FM 23-67).

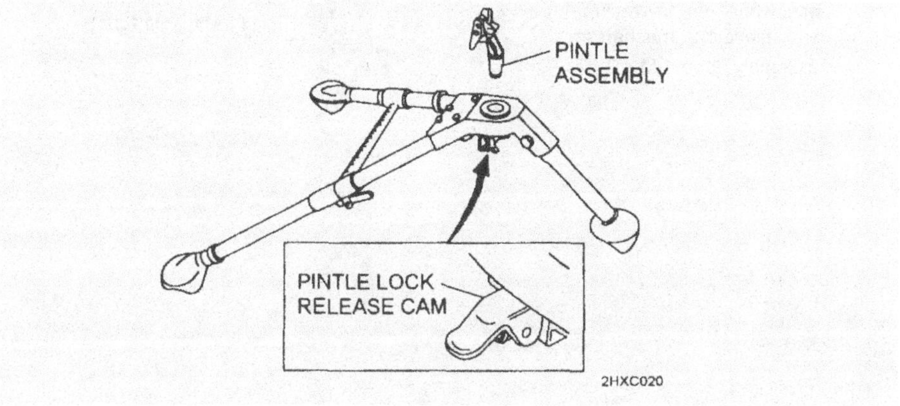

OPERATION UNDER USUAL CONDITIONS - HOW TO PUT THE M60 ON THE M122 TRIPOD (cont) 0006 00

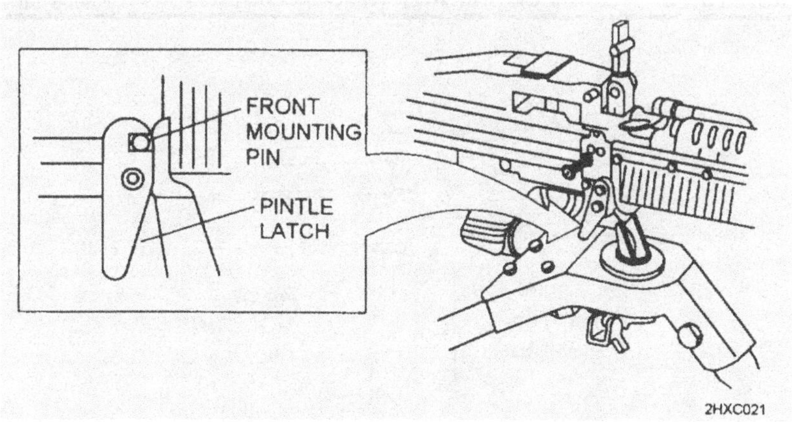

3 Place ends of machine gun's front mounting pin on top of pintle assembly.

4 Press bottom of latch to open pintle assembly. The ends of machine gun mounting pin should lock in place on pintle assembly.

5 Attach gun mount adapter to Traversing and Elevating Mechanism using straight pin and clips.

6 Center (zero) traversing hand-wheel and elevating handwheel on traversing and elevating mechanism assembly (see FM 23-67).

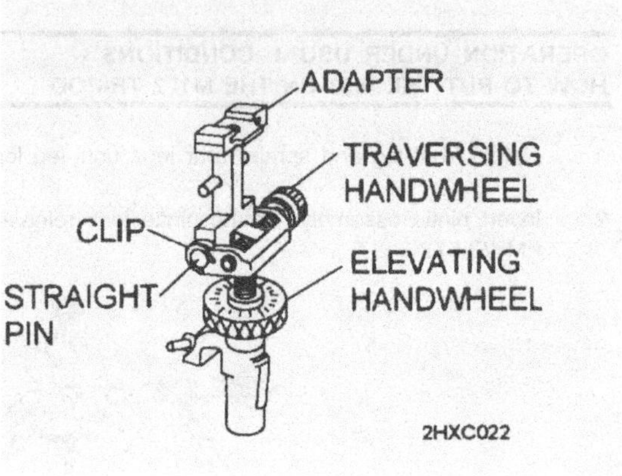

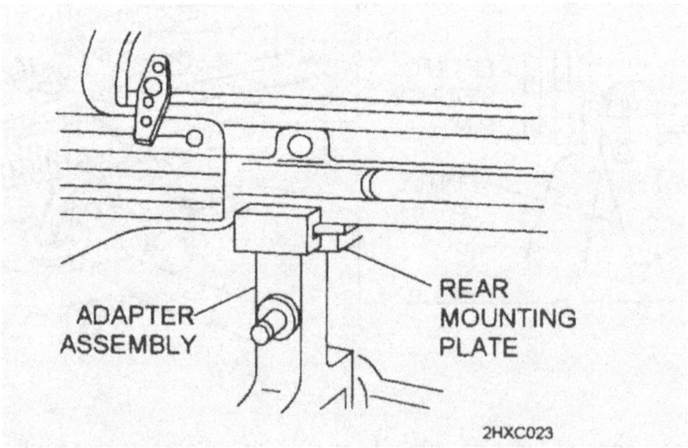

7 Slide adapter assembly on machine gun's rear mounting plate from the rear to the front.

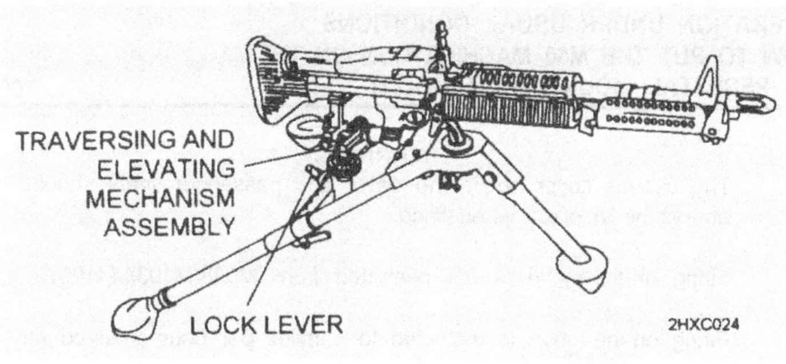

8 Position traversing and elevating mechanism assembly on traversing bar. Rotate lock lever to secure.

OPERATION UNDER USUAL CONDITIONS - HOW TO PUT THE M60 MACHINE GUN ON THE M6 PEDESTAL MOUNT/ M197 MOUNT 0007 00

NOTE

For mounting instructions see TM 9-1005-245-13&P.

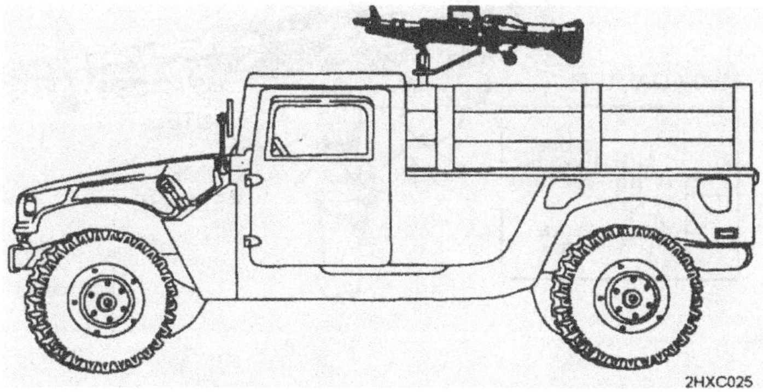

WARNING

The canvas cover above the driver and passenger seats should always be in place when firing.

Firing on-the-move is not permitted from M998/M1038/M1097.

Firing on-the-move is restricted to 5 miles per hour cross-country and 10 miles per hour on improved roads when mounted on M1025 and M1026 HMMWVs (Armament Carriers).

Ground personnel should not be within 10 meters of the vehicle when firing.

CAUTION

When the weapon is not in use/manned it should be secured in the travel lock. When not manned for extended cross-country travel weapon should be removed from the mount,

1 Ensure bipod legs are in the rearward locked position and the M197 pintle latch is rearward.

2 Press bottom of latch to open the pintle assembly. The ends of the machine gun mounting pin should lock in place on the pintle assembly.

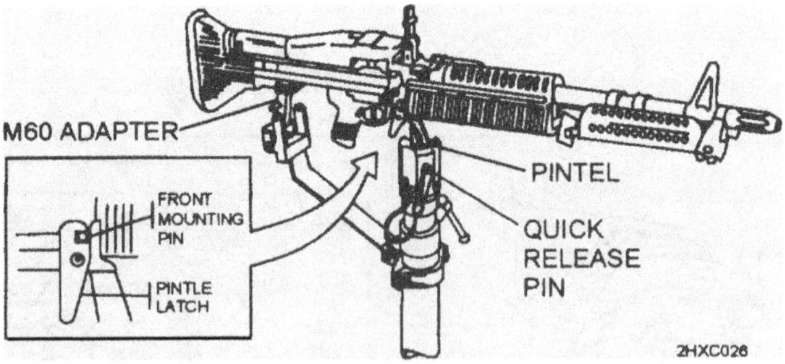

3 Place the end of the machine gun's front mounting pin on top of the pintle assembly.

4 Pivot the M60 adapter assembly into the upright position.

5 Slide the adapter assembly on the machine gun's mounting plate from the rear to the front.

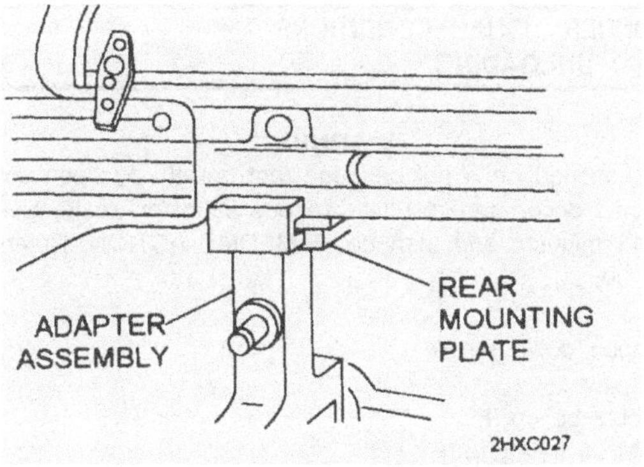

OPERATION UNDER USUAL CONDITIONS - HOW TO PUT M60D ON VEHICLES 0008 00

1 Pull quick release pin.

2 Lower gun adapter.

3 Attach M60D machine gun to aircraft mount using applicable manual.

4 Raise gun adapter.

5 Insert quick release pin

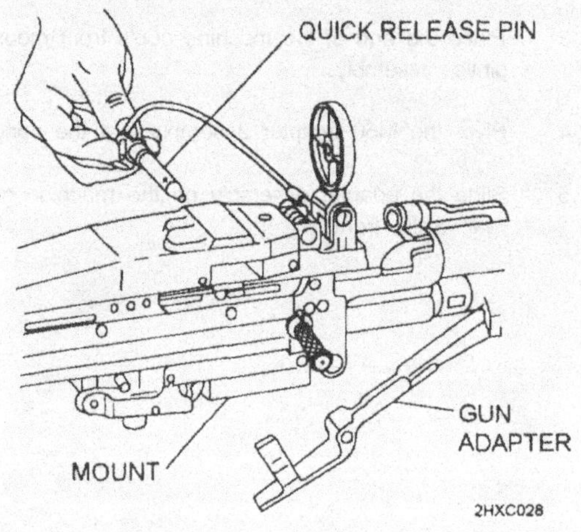

OPERATION UNDER USUAL CONDITIONS - CLEARING AND UNLOADING 0009 00

WARNING

If cover is opened on a hot cartridge (hot barrel), an open cover cookoff could occur and result in serious injury or death. Evacuate area for 15 minutes and then do REMEDIAL ACTION. Refer to WP 0011 00.

1 Point weapon down range.

2 Safety should be on 'F'.

- M60 - Move safety to 'F'.
- M60D - Press in on 'F'.

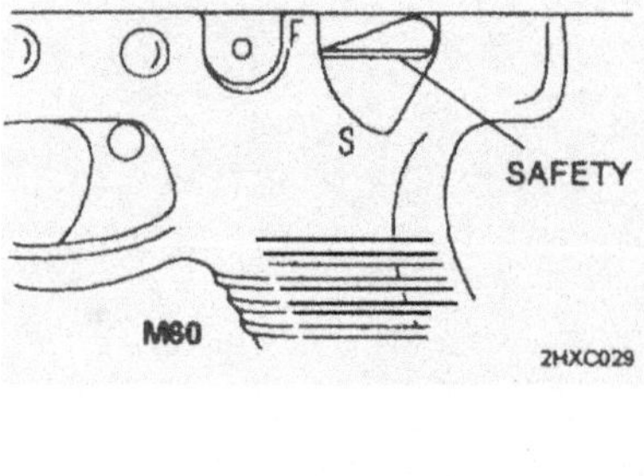

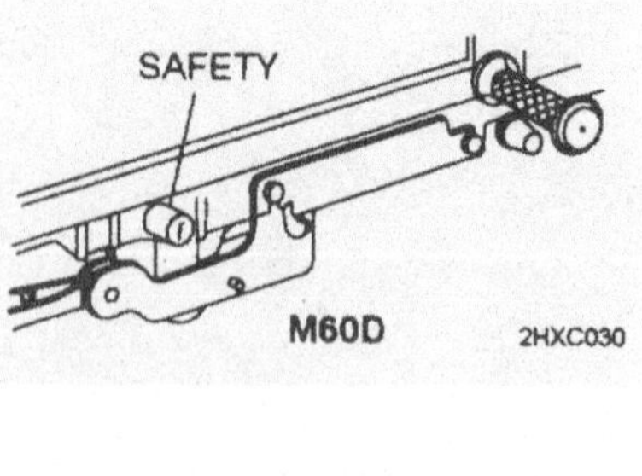

3 Pull cocking handle rearward, then push it forward until it locks.

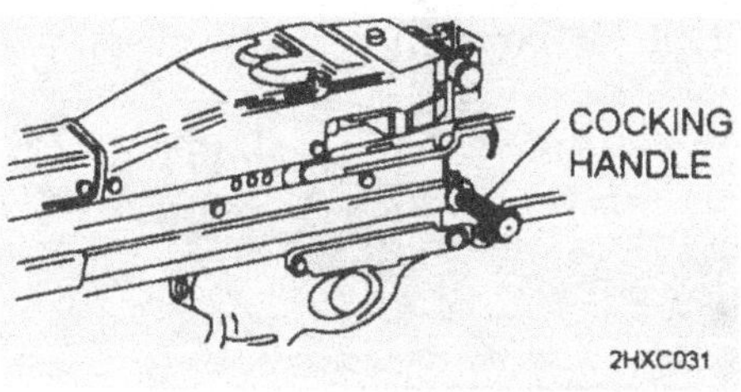

4 M60 - Move safety to 'S'.
M60D - Press in on 'S'.

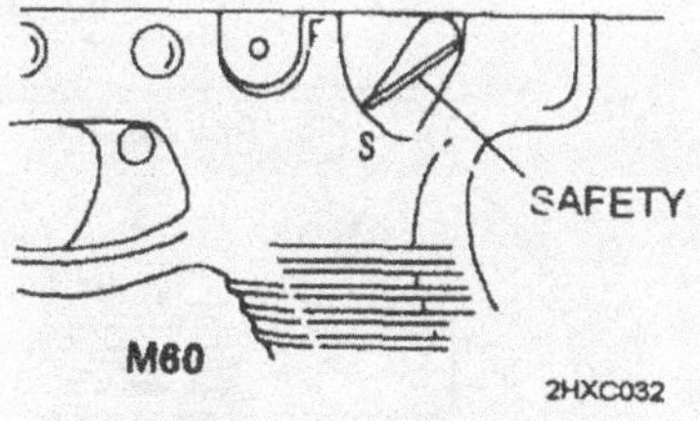

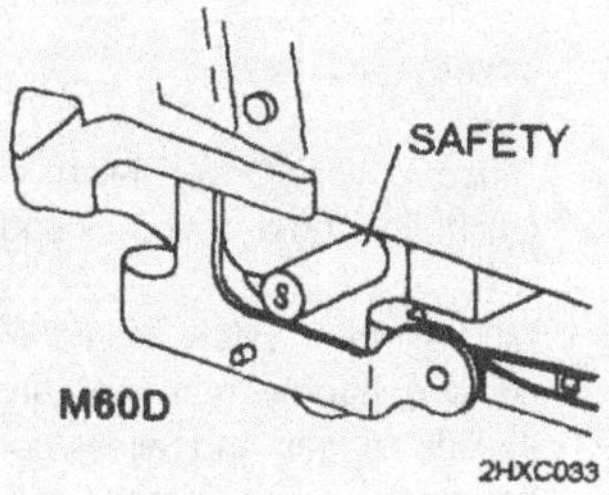

5 Turn latch lever and open cover.

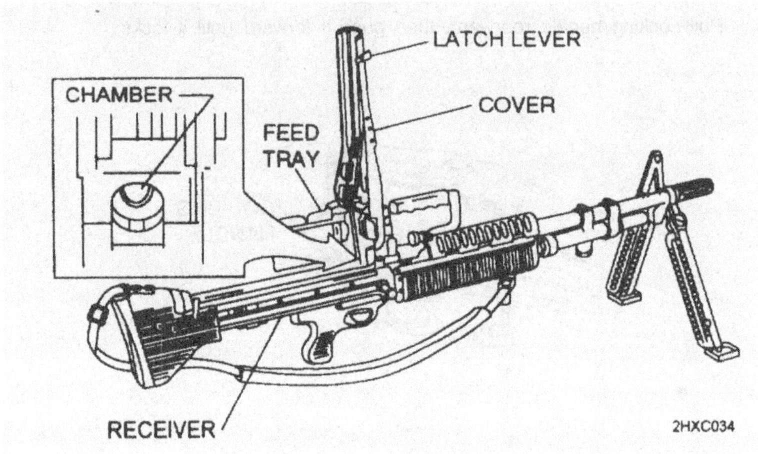

6 Remove ammunition and link belt and raise feed tray.

7 Eyeball chamber. If empty, go to step 8. If a round is chambered, see WP 0012 00 for stuck or ruptured cartridge case and WP 0017 00 for troubleshooting.

8 Lower feed tray.

9 Close cover. Cover latch should catch and hold cover closed.

10 M60 - Move safety to 'F'. Pull cocking handle rearward and hold it. Pull trigger and allow cocking handle to move forward slowly.

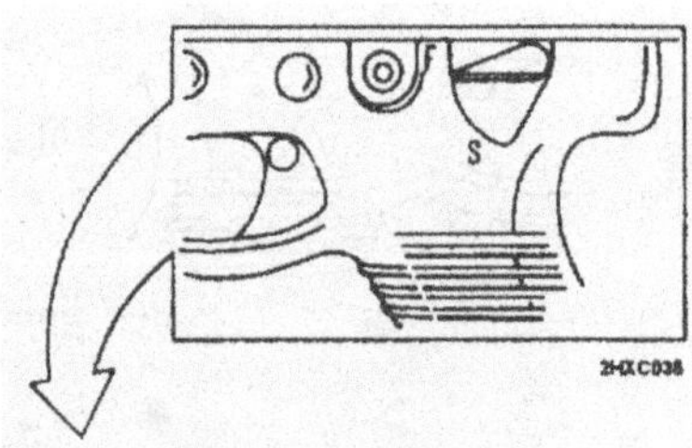

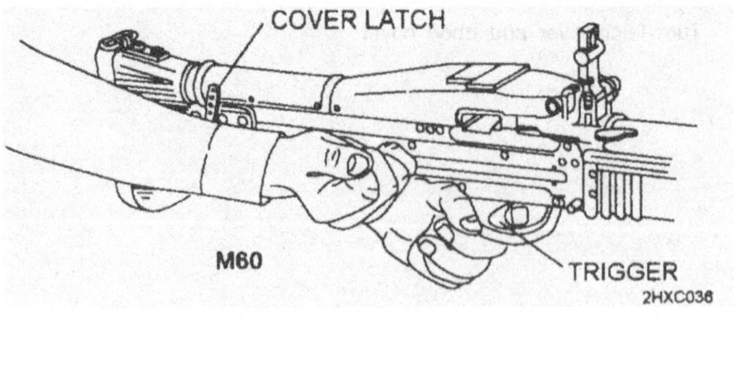

M60D - Press in on 'F'. Pull cocking handle rearward and hold it. Pull trigger and allow cocking handle to move forward slowly.

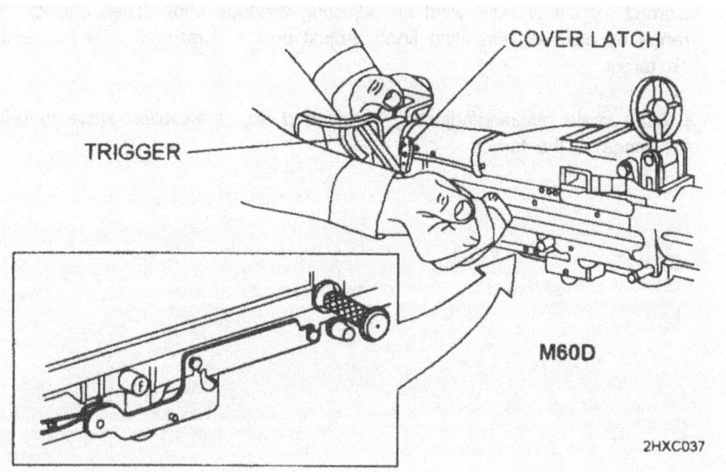

INITIAL ADJUSTMENTS, DAILY CHECKS, AND SELF-TEST 0010 00

FIELD ZEROING

NOTE

One mil is equal to one meter at 1000 meters.

1 Choose a target between 300 and 700 meters.

2 Place range scale to target on rear sights.

3 Zero windage scale with index line.

4 Aim at 6 o'clock on bottom of target.

5 Fire a burst of 6 to 9 rounds and observe impact of burst.

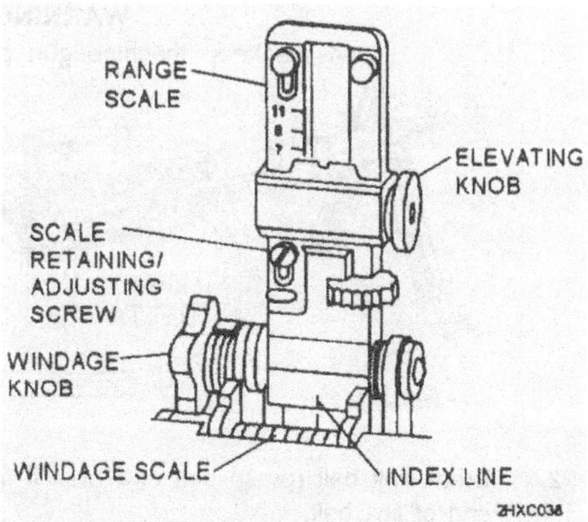

FIELD ZEROING (cont)

6 Correct for left or right wind by adjusting windage knob. Then correct range by adjusting elevating knob. Adjust until a burst of 6 to 9 rounds is on target.

7 Loosen scale retaining/adjusting screw and adjust elevation scale to reflect the range to the target.

OPERATING PROCEDURES 0011 00

IMMEDIATE ACTION

RUNAWAY M60 MACHINE GUN

1 If runaway occurs (machine gun won't stop firing), take the following actions to correct it quickly.

WARNING

Always keep machine gun pointed down range.

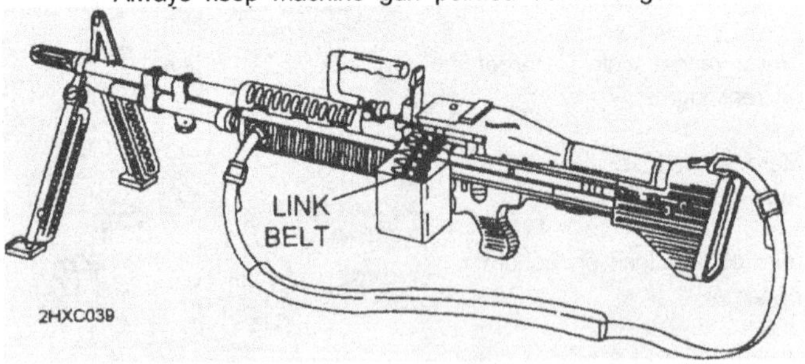

2 Break link belt (grasp belt and twist it firmly) or let machine gun fire if near end of link belt.

IMMEDIATE ACTION (cont)

RUNAWAY M60 MACHINE GUN (cont)

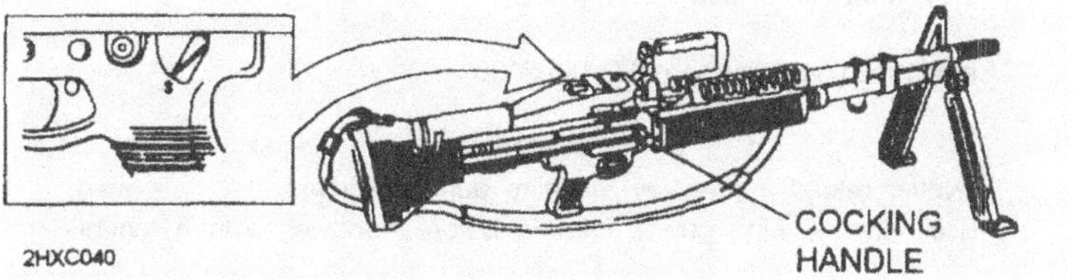

3 Pull cocking handle all the way back and hold it. Place safety to 'S' and remove link belt.

4 Clear machine gun (see WP 0009 00).

WARNING

Never reload a runaway machine gun until it has been repaired. Be sure machine gun is cleared before moving it.

5 Notify unit maintenance for repair.

RUNAWAY M60D MACHINE GUN

1 If runaway occurs (machine gun won't stop firing), take the following actions to correct it quickly.

WARNING

Always keep machine gun pointed down range.

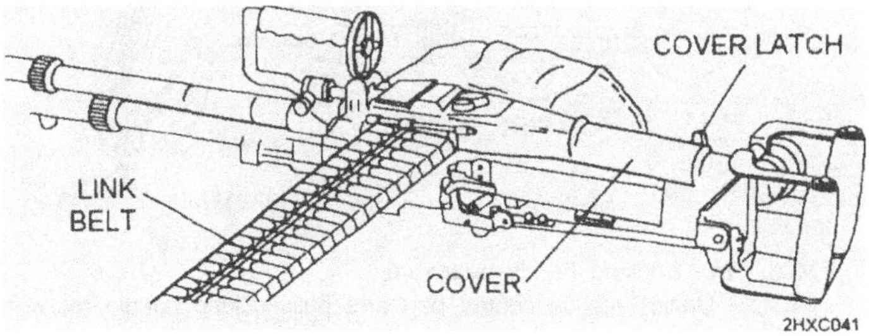

2 Let machine gun fire if near end of link belt.

3 Rotate cover latch and open cover.

IMMEDIATE ACTION (cont)

RUNAWAY M60 MACHINE GUN (cont)

4 Clear machine gun (see WP 0009 00).

WARNING

Never reload a runaway machine gun until it has been repaired. Be sure machine gun is cleared before removing it from mount.

5 Notify unit maintenance for repairs.

MISFIRE

If the weapon stops firing before you complete your mission, take IMMEDIATE ACTION within 10 seconds.

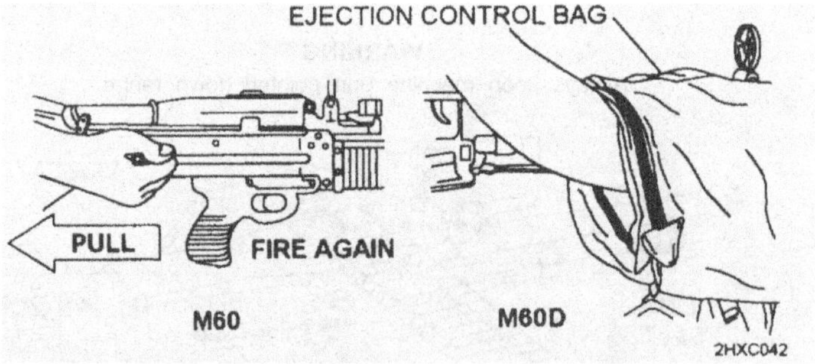

1 M60 - Pull cocking handle rearward.
M60D - Unzip ejection control bag and pull cocking handle rearward.

2 Ensure bolt is locked rearward and return cocking handle to forward locked position.

3 Check ejection port (M60 only).

MISFIRE (cont)

4 If a round or cartridge case is ejected, attempt to fire again.

WARNING

If cover is opened on a hot cartridge (hot barrel), an open-cover cookoff could occur and result in serious injury or death. Close cover and evacuate area for 15 minutes.

The climate temperature in different regions will make a difference as to what constitutes a hot gun. A hot, sunny day can cause a cookoff within 50 rounds, weapon and ammunition in sun.

5 If nothing is ejected and the barrel is hot enough to cause a cookoff (200 rounds fired within 2 minutes), watt at least 15 minutes (make sure bolt is locked rearward). Repeat steps for IMMEDIATE ACTION.

6 If IMMEDIATE ACTION fails to remove cartridge case, take REMEDIAL ACTION.

REMEDIAL ACTION

STEPS TO TAKE IF IMMEDIATE ACTION - MISFIRE - DOES NOT WORK

WARNING

If cover is opened on a hot cartridge (hot barrel), an open-cover cookoff could occur and result in serious injury or death. Close cover and evacuate area for 15 minutes.

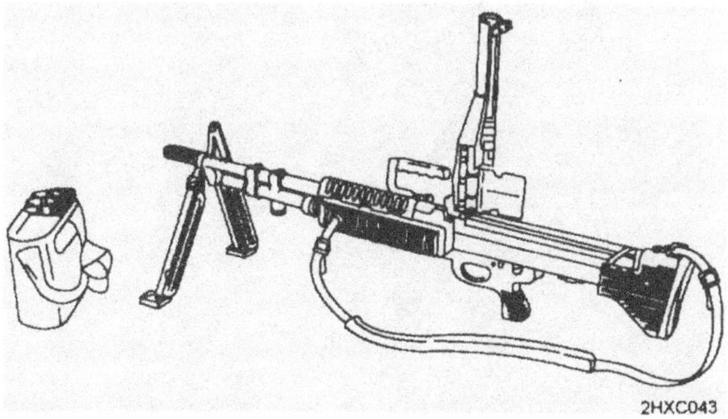

REMEDIAL ACTION (cont)

1 Keep weapon on target (down range/impact area). Clear weapon when barrel is cool (after 15 minutes wait) using the following steps.

2 Pull cocking handle rearward locking bolt to the rear. Return charging handle to forward locked position.

3 If a round is not ejected:
M60 - Place safety to 'S'.
M60D - Press in on 'S'.

4 Open cover, remove ammunition link belt and raise feed tray.

5 Inspect receiver, chamber, ejector, and ammunition.

6 If a round is in chamber, lower feed tray and close cover.

7 M60 - Place safety to 'F'.
M60D - Press in on 'F'.

6 Attempt to fire.

9 If a round is fired and ejected, reload and continue to fire.

10 If weapon does not fire or eject, clear and unload weapon, and notify unit maintenance for repair.

WARNING

When removing a stuck unfired cartridge, stay clear of the muzzle. Do not allow cartridge to contact any hard surface. Cartridge may fire on contact. Remove a stuck unfired cartridge using the same procedures for removing a stuck or ruptured cartridge case.

1 Pull cocking handle rearward and place safety to 'S'.

2 Remove barrel (see WP 0018 00).

3 **STUCK CARTRIDGE CASE/ UNFIRED CARTRIDGE-** Insert cleaning rod in muzzle and tap out case/cartridge.

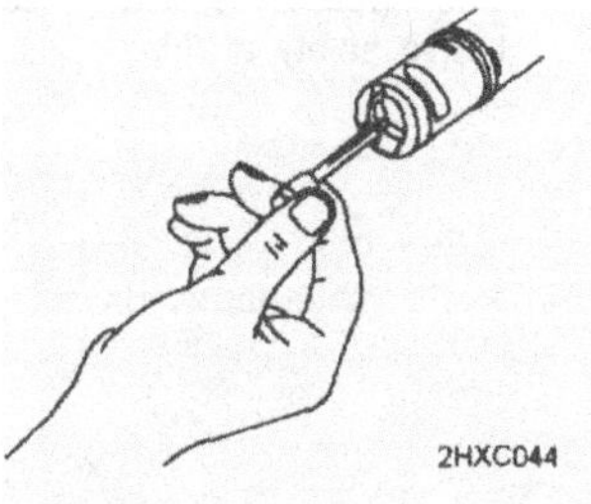

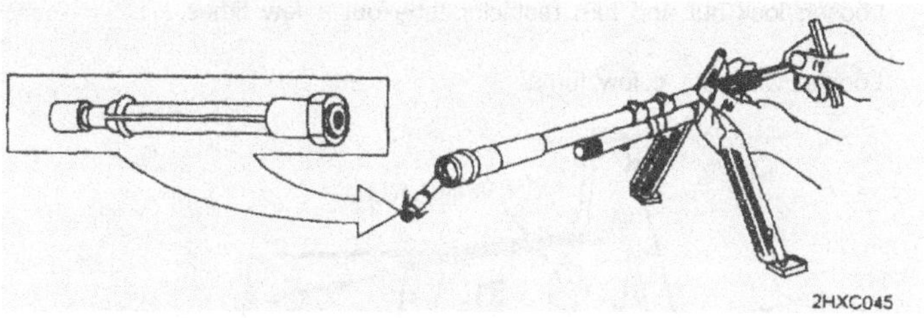

3 **RUPTURED CARTRIDGE CASE (cont)**

a. Push ruptured cartridge extractor inside ruptured case in chamber until it seats.

b. Lightly tap out with cleaning rod.

c. Remove extractor from cartridge case.

d. Unscrew extractor tool and remove cartridge case. Reassemble extractor tool.

4 Install barrel (see WP 0018 00).

5 Close cover.

6 Hold cocking handle rearward.

7 Place safety to ‘F’.

8 Pull trigger and ease cocking handle forward.

9 Once weapon is cleared, do not place safety to ‘S’ when bolt is forward. Leave safety at ‘F’.

OPERATION OF AUXILIARY EQUIPMENT 0013 00

INSTALLING M13A1 BLANK FIRING ATTACHMENT

1 Loosen lock nut and turn restrictor tube out a few times.

2 Loosen wing nut a few turns.

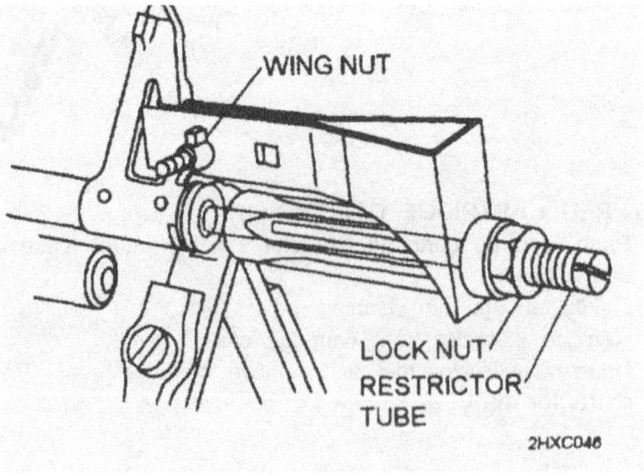

INSTALLING M13A1 BLANK FIRING ATTACHMENT (cont)

3 Slide restrictor tube in flash suppressor as far as possible,

4 Clamp blank firing attachment around front sight.

5 Tighten wing nut fingertight.

6 Screw restrictor tube in until it seats tight against muzzle end of barrel to prevent gas leakage.

7 Tighten lock nut.

NOTE

After Firing 500 Blank Rounds: Clean gun including barrel assembly, gas cylinder, gas piston, gas port, chamber bore, and Blank Firing Attachment (BFA).

NIGHT VISION SIGHT MOUNTED ON M60 MACHINE GUN

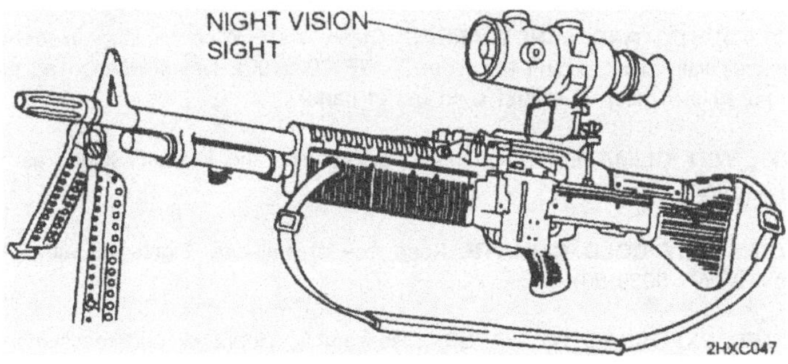

To mount night vision sight AN/PVS-4 on the M60 machine gun, refer to TM 11-5655-213-10.

PREPARATION FOR MOVEMENT 0014 00

Refer to FM 23-67.

OPERATION UNDER UNUSUAL CONDITIONS 0015 00

Under unusual conditions, clean and lubricate machine gun more often.

HOT, DUSTY, AND SANDY AREAS. Clean often. Wipe oil from exposed surfaces with clean wiping rag (item 5, WP 0039 00). Cover weapon as much as possible to keep dust and sand out of parts.

HOT, WET CLIMATE. Inspect often. Dry, clean, and lubricate lightly as necessary.

EXTREMELY COLD CLIMATE. Keep free of moisture. Lightly oil with LAW (item 3, WP 0039 00).

AFTER EXPOSURE TO WATER. Disassemble, clean, oil and reassemble as soon as possible. Make sure it's dry.

NUCLEAR, BIOLOGICAL, AND CHEMICAL (NBC) DECONTAMINATION. Decontamination procedures can be found in FM 3-5.

CHAPTER 3

MAINTENANCE INSTRUCTIONS

LUBRICATION INSTRUCTIONS 0016 00

LUBE GUIDE

Under all but the coldest arctic conditions, LSA (item 4, WP 0039 00), and CLP (item 1, WP 0039 00) are the lubricants to use on your machine gun. Remember to remove excessive oil from the bore before firing.

- Lightly Lubricate. A film of lubricant barely visible to the eye.
- Generously Lubricate. Heavy enough so lubricant can be spread with finger.

NOTE

Lubrication instructions are mandatory. Do not mix lubricants on the same machine gun. The weapon must be thoroughly cleaned during change from one lubricant to another. Dry cleaning solvent (available to unit maintenance) is recommended for cleaning during change from one lubricant to another.

LUBE GUIDE (cont)

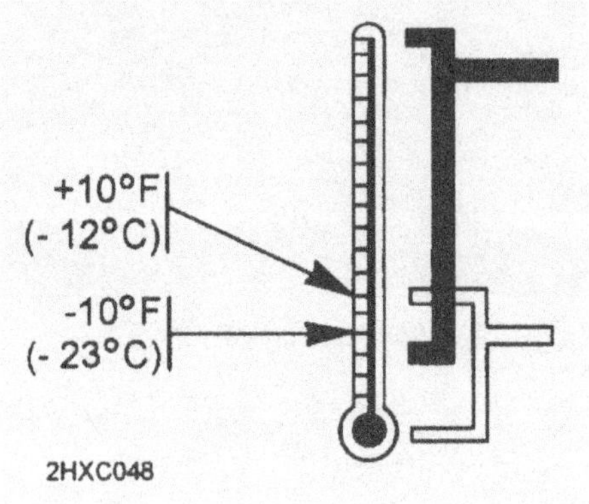

CLP - Cleaner, lubricant and preservative (item 1, WP 0039 00)

LSA - Weapons lubricating oil, semi-fluid (item 4, WP 0039 00)

LAW - Weapons lubricating oil, arctic (item 3, WP 0039 00)

Between 10°F (-12°C) and -10°F (-23°C) use CLP, LSA or LAW.
Below -10°F (-23°C) use only LAW.

WARNING

Be sure to clear weapon before disassembling, cleaning, inspecting, or storing.

1 After firing, field-strip your machine gun (WP 0018 00). Clean bore and chamber with cleaning rod, chamber brush, bore brush, and swab soaked with rifle bore cleaner (RBC) or CLP until a clean swab can be run through the bore without getting dirty.

2 Clean powder-fouled parts, except the buffer, with a wiping rag dampened with RBC or CLP.

3 Wipe dry and lube as required. Inspect and run a lightly oiled swab through the bore and chamber.

4 If your machine gun is not used, it still needs complete cleaning and lubing at least every 90 days. (Unusual conditions could shorten this time.)

LUBE GUIDE (cont)

5 External Surfaces. Put CLP on a clean swab and generously lubricate all shaded areas. Lightly lubricate all other areas except rubber-coated parts.

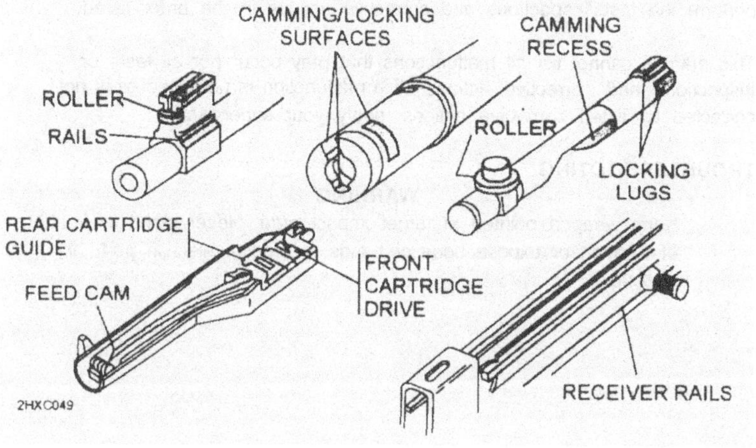

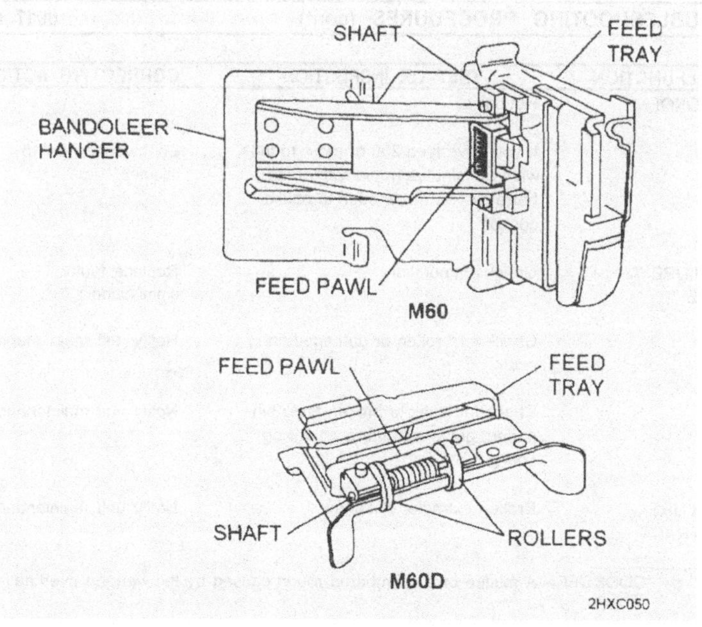

INTRODUCTION

The table lists the common malfunctions which you may find during the operation or maintenance of the machine gun or its components. You should perform the tests/inspections and corrective actions in the order listed.

This manual cannot list all malfunctions that may occur nor all tests or inspections and corrective actions. If a malfunction is not listed or is not corrected by listed corrective actions, notify your supervisor.

TROUBLESHOOTING

WARNING

Keep weapon pointed at target impact area. Never stand in front of weapon or expose body or hands to breech, ejection port, or muzzle.

MALFUNCTION	TEST OR INSPECTION	CORRECTIVE ACTION
COOKOFF*	Hot barrel	
	If you have tired 200 or more rounds within 2 minutes (rapid rate of tire), barrel will be hot enough to cause cookoff.	Cool weapon for 15 minutes.
FAILURE TO FIRE	Check ammunition.	Replace faulty ammunition.
	Check for broken or damaged firing pin.	Notify unit maintenance
	Check for weak or broken firing pin spring, guide assembly, or driving spring	Notify unit maintenance
FAILURE TO EXTRACT	Broken extractor or spring.	Notify unit maintenance

*COOKOFF - A misfire of a chambered round caused by the weapon overheating

MALFUNCTION	TEST OR INSPECTION	CORRECTIVE ACTION
FAILURE TO EXTRACT (cont)	Short recoil.	Clean gas port, piston, gas cylinder, operation rod assembly, and chamber (see WP 0023 00).
	Gas piston installed backwards	Install correctly.
	Dirty ammunition or chamber.	Clean chamber and/or use new ammunition.
SLUGGISH OPERATION	Friction from dirt, carbon, burrs or lack of lubrication.	Clean and lubricate or notify unit maintenance of any burrs.
WADDING AND UNBURNED POWDER	Manually charge weapon (if firing blanks)	Clean and lubricate or notify unit maintenance.
FAILURE TO CHAMBER	Ruptured cartridge case	Remove cartridge case (WP 00012 00).
	Carbon buildup in gas cylinder	Remove carbon.
	Carbon buildup in receiver assembly.	Remove carbon

MALFUNCTION	TEST OR INSPECTION	CORRECTIVE ACTION
FAILURE TO CHAMBER (cont)	Damaged round.	*Remove round and recharge weapon.
	Dirty chamber.	Clean or replace barrel.
FAILURE TO EJECT	Frozen, damaged or missing ejector spring.	Notify unit maintenance.
	Short recoil.	Clean gas port, piston, gas cylinder, chamber, and operating rod guide (WP 0023 00).
FAILURE TO COCK OR RUNAWAY GUN	Operating rod sear notch or sear worn.	Notify unit maintenance
	Sear plunger or spring broken or missing.	Notify unit maintenance.

*Dispose of damaged round according to range SOP

MALFUNCTION	TEST OR INSPECTION	CORRECTIVE ACTION
FAILURE TO FEED	Insufficient gas pressure (short recoil).	Clean gas system, chamber, and barrel socket (WP 0023 00).
	Improper lubrication.	Lubricate as necessary (WP 0016 00).
	Defective ammunition or links.	Use new ammunition belt.
	Ammunition belt installed wrong.	Install with open side of link down (WP 0005 00).
	Damaged or weak operating parts.	Notify unit maintenance
	Obstruction in receiver assembly.	Remove

END OF TASK

FIELD-STRIPPING MACHINE GUN 0018 00

M60 SHOULDER GUN STOCK

WARNING

Before field stripping, clear the weapon and check that the bolt is in forward position.

NOTE

The following procedures illustrate the M60 machine gun except where M60 parts differ from M60D pans, then both weapons are illustrated.

1 Lift shoulder rest (1), insert cleaning rod (2) in hole to release latch. Pull shoulder gun stock (3) from receiver.

2 Turn latch lever (4) and open cover.

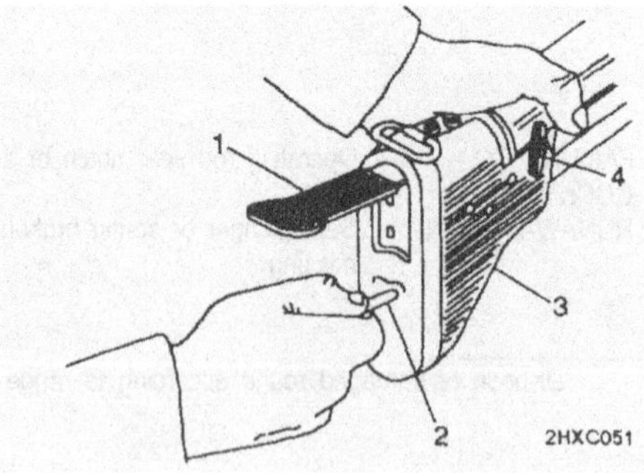

M60D SPADE AND TRIGGER GRIP ASSEMBLY

WARNING

To prevent possible bodily injury and aircraft damage, personnel should not stand the weapon on its barrel assembly when disassembling or assembling the weapon.

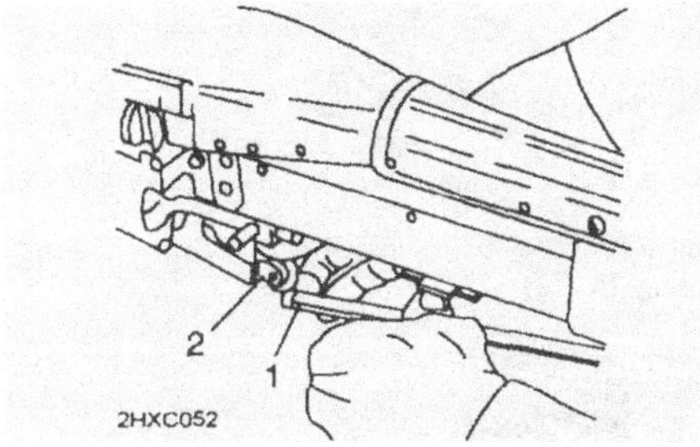

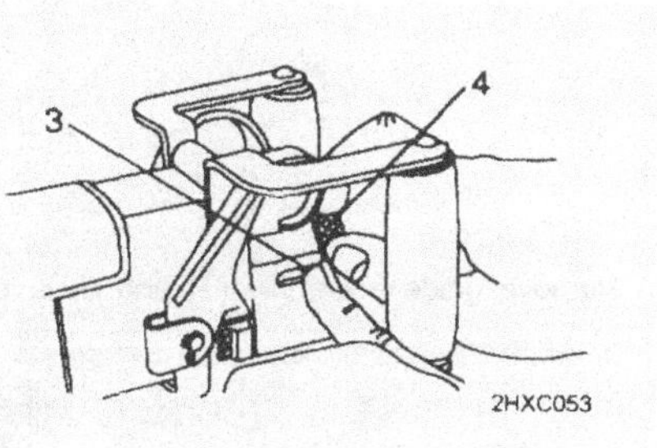

1 Unsnap sear assembly link and spring (1) from rear of sear assembly activator (2) by pressing upward with fingers.

2 Press and hold end of spring pin (3) forward while unscrewing knob and pin assembly (4).

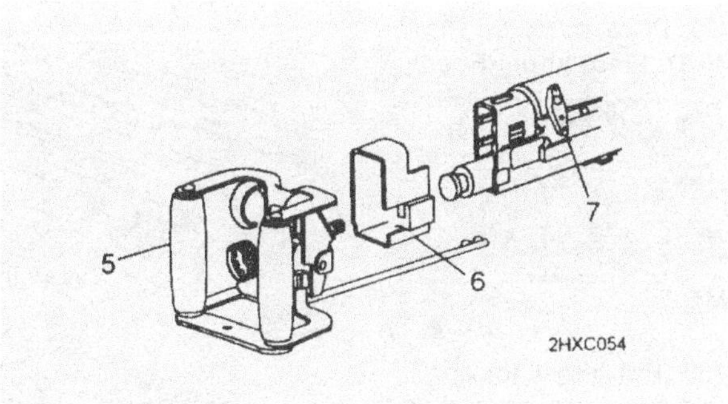

3 Remove grip and trigger assembly (5) and dust and moisture boot (6).

4 Turn latch lever (7) and open cover.

BOLT ASSEMBLY AND OPERATING ROD ASSEMBLY

1 Apply slight palm pressure to rear of hydraulic buffer assembly (1). Lift out buffer retaining yoke (2) and remove buffer assembly (1).

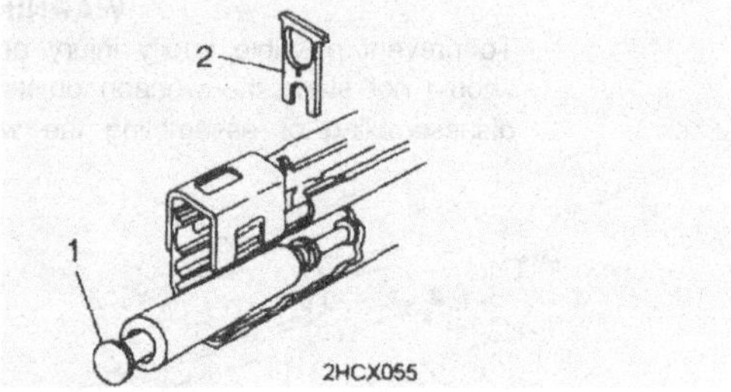

2 Remove guide assembly (3) and drive spring (4).

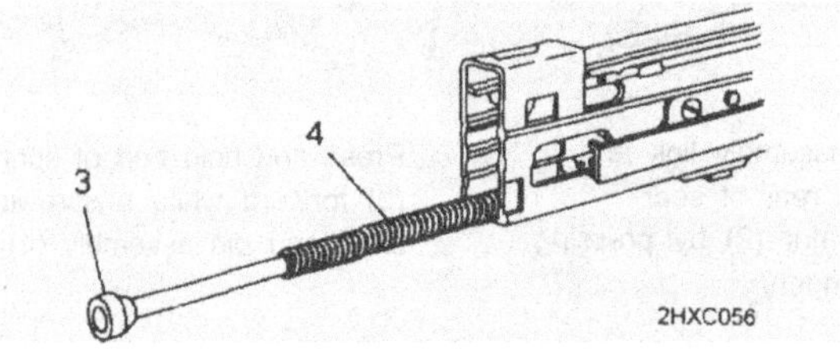

3 M60. Place safety to ‘F’. M60D. Press in on ‘F’.

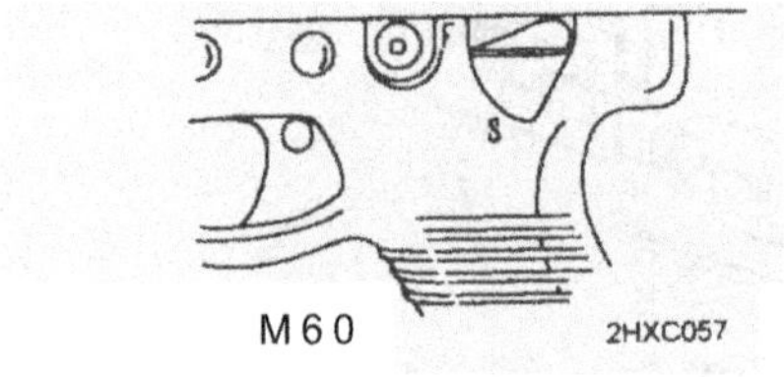

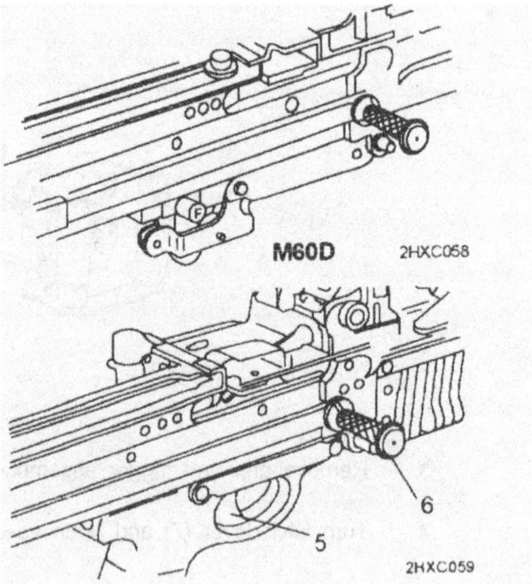

4 M60. Pull trigger (5) and hold it while pulling cocking handle assembly (6) rearward.

BOLT ASSEMBLY AND OPERATING ROD ASSEMBLY (cont)

M60 D. Pull down and hold sear assembly activator (7) while pulling cocking handle assembly (6) rearward.

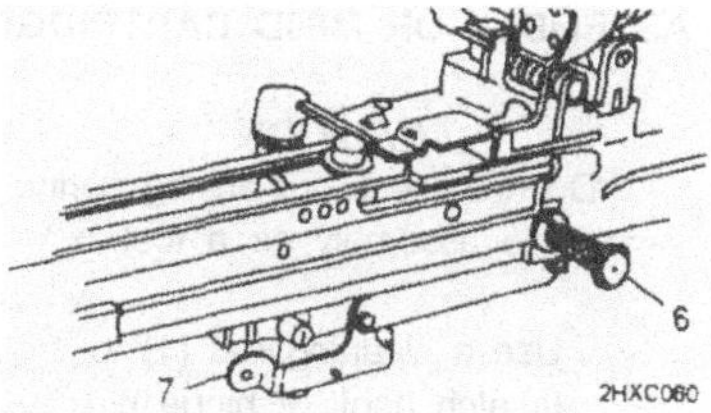

WARNING

Bolt assembly is under spring tension; it can twist and injure your hand.

5 Removing operating rod assembly (8) and bolt assembly (9) as a unit.

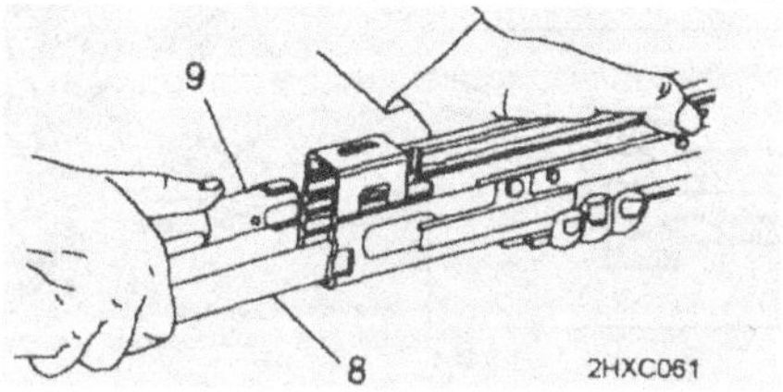

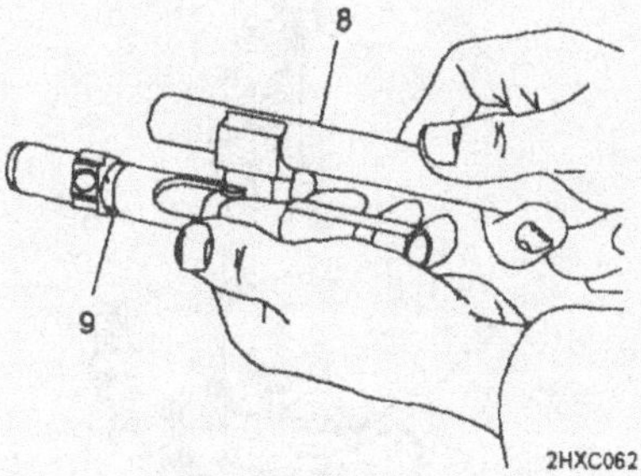

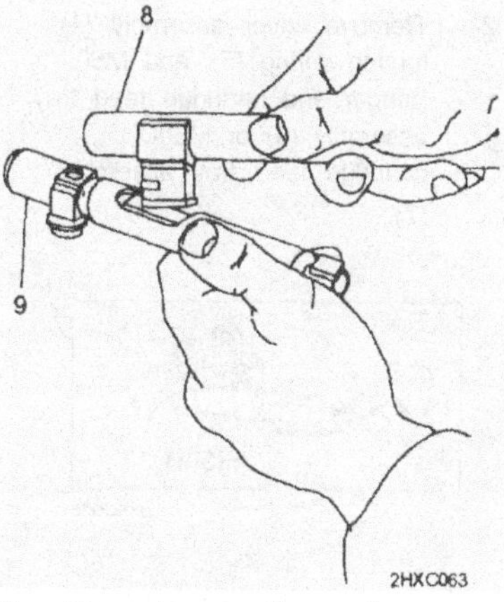

6 Push operating rod assembly (8) toward rear of slot in bolt assembly (9).

7 Separate operating rod assembly (8) from bolt assembly (9).

COVER ASSEMBLY AND M60 HANGER AND CARTRIDGE FEED TRAY ASSEMBLY OR M60D CARTRIDGE FEED TRAY ASSEMBLY

CAUTION

Do not use tip of driving spring guide assembly as a tool.

Use a cleaning rod (1) to unlatch hook of hinge pin latch (2). Remove hinge pin latch (2) and cover hinge pin (3).

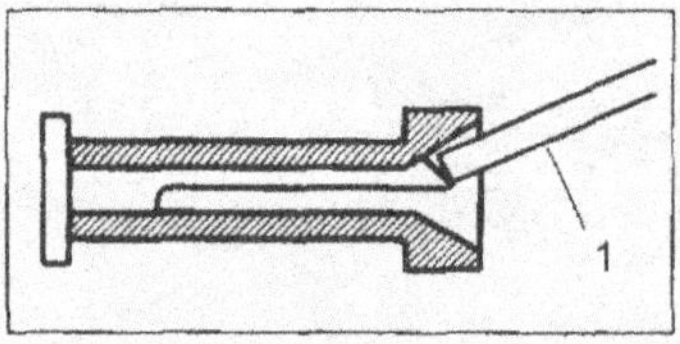

2HXC064

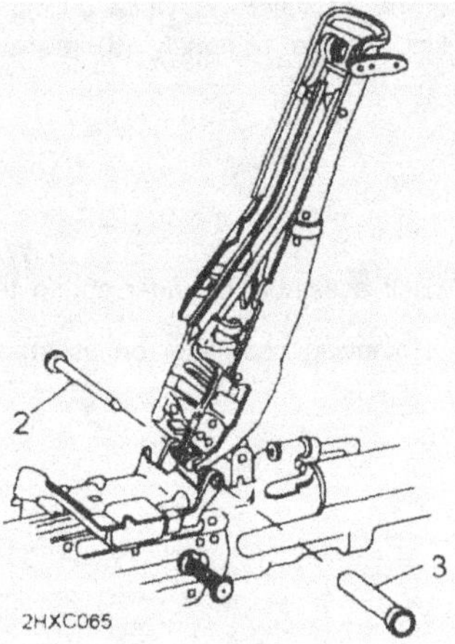

2HXC065

2 Remove cover assembly (4), torsion spring (5), and M60 hanger and cartridge feed tray assembly (6) or M60D cartridge feed tray assembly (7).

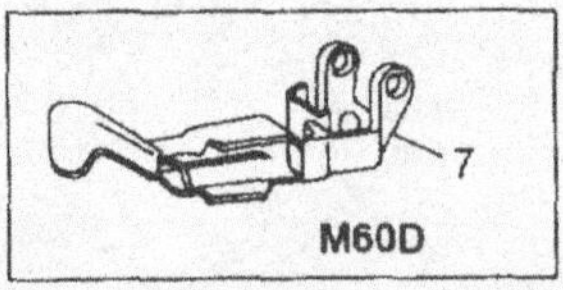

2HXC066

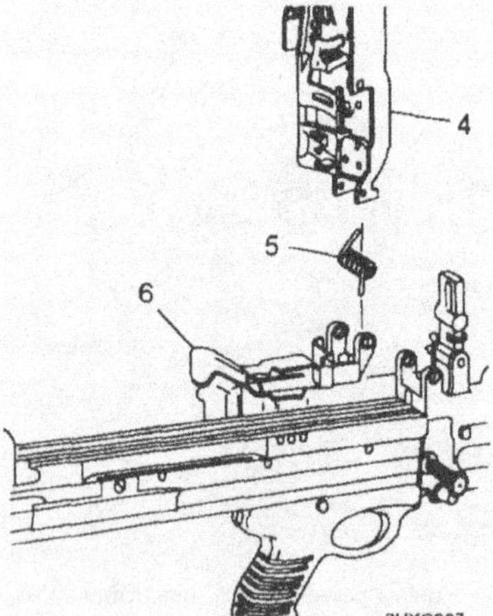

2HXC067

BARREL AND BIPOD ASSEMBLY

1 M60D. Pull quick release pin (4) out of receiver assembly and let gun adapter (5) swing down.

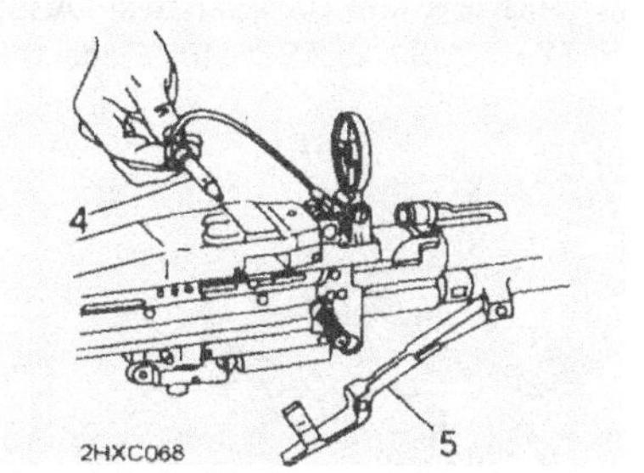

2 Push in spring detent (1) and raise barrel lock (2). Remove barrel and bipod assembly (3).

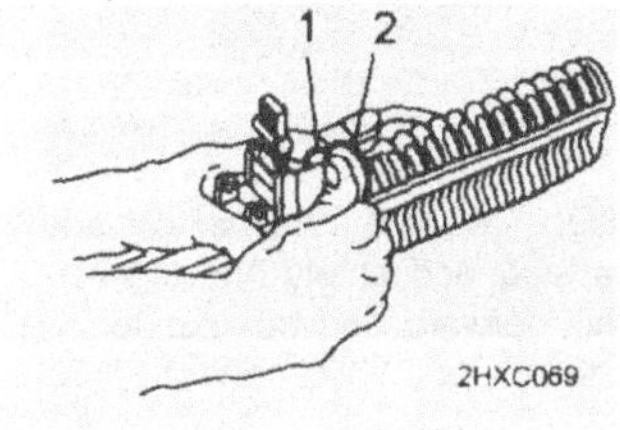

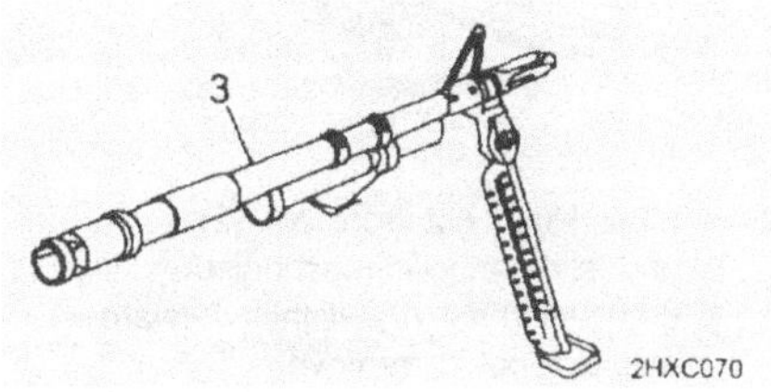

M60 TRIGGER MECHANISM GRIP ASSEMBLY OR M60D SEAR AND SAFETY HOUSING ASSEMBLY

CAUTION

Care should be taken, when disassembling M60 and M60D in the same work area, to ensure proper leaf spring remains with the appropriate model weapon.

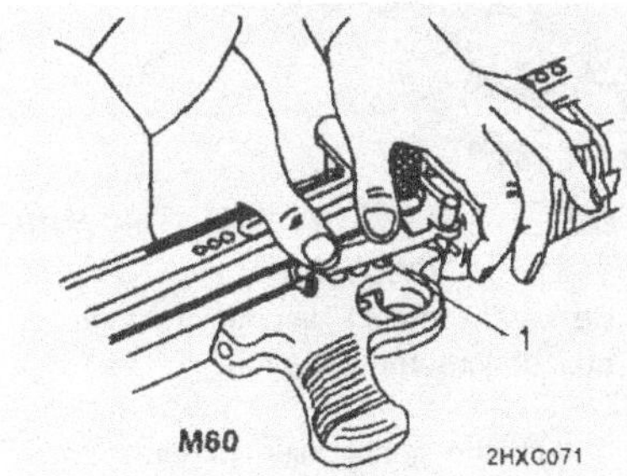

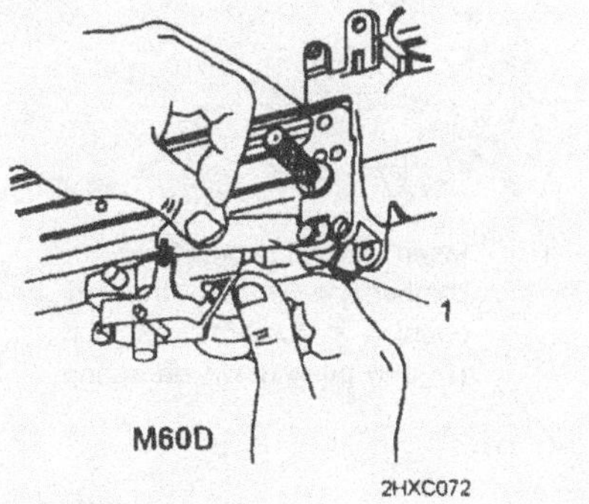

M60 and M60D. Push in and remove flat leaf spring (1).

M60 TRIGGER MECHANISM GRIP ASSEMBLY OR M60D SEAR AND SAFETY HOUSING ASSEMBLY (cont)

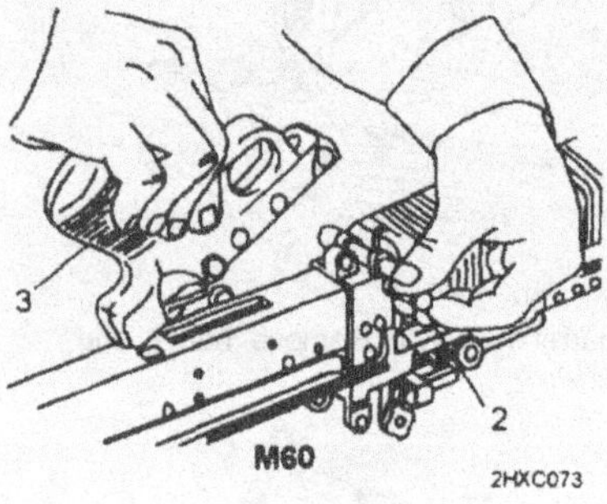

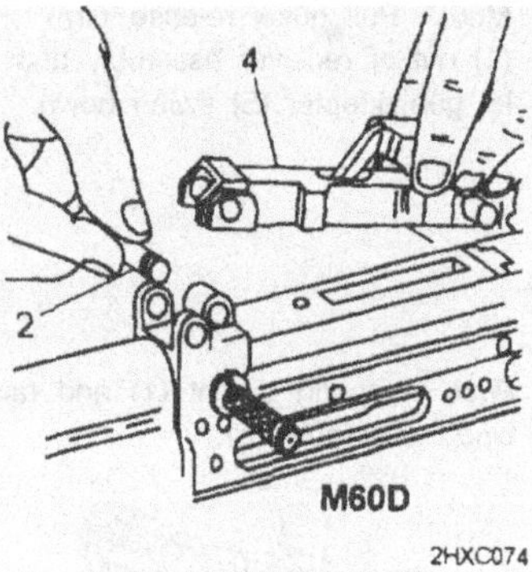

2 M60. Push out front pin (2) and slide trigger mechanism grip assembly (3) slightly forward and pull out to remove.

M60D. Push out front pin (2) and slide sear and safety housing (4) slightly forward and pull out to remove.

FOREARM ASSEMBLY

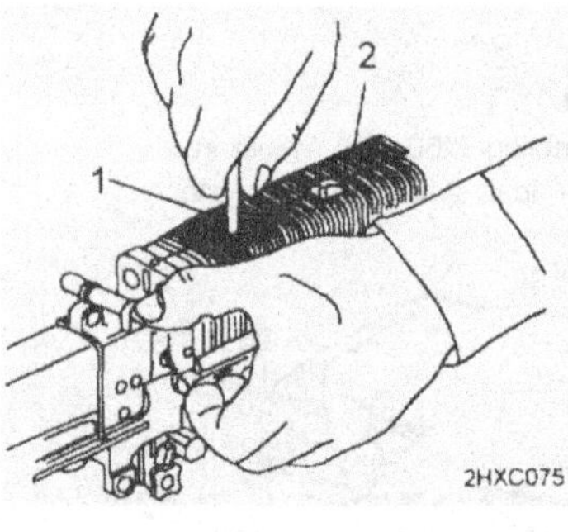

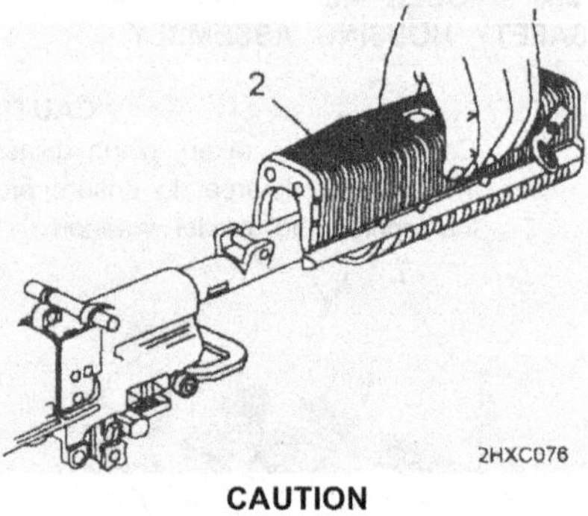

1 Insert cleaning rod (1) or combination wrench through opening in forearm assembly (2) and push down on spring.

CAUTION

Be careful not to damage internal ribs of forearm assembly.

2 Lift and gently slide forearm assembly (2) from receiver.

END OF TASK

MAINTENANCE OF BOLT ASSEMBLY 0019 00

INSPECT AND CLEAN

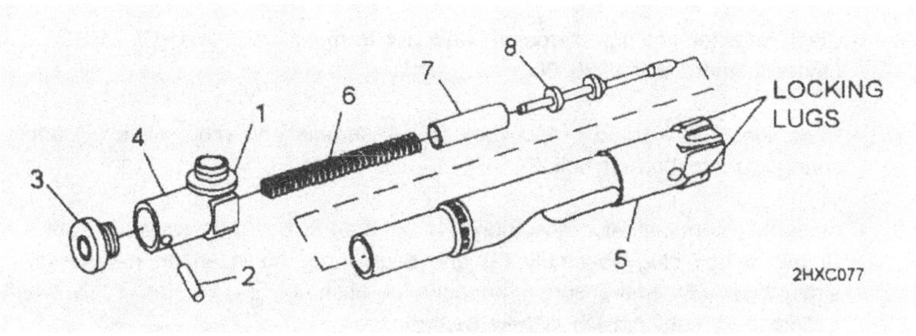

1 Inspect bolt assembly for burrs and cracks, especially in the locking lug area.

2 Make sure roller (1) operates freely and is not cracked.

MAINTENANCE OF BOLT ASSEMBLY (cont) 0019 00

INSPECT AND CLEAN (cont)

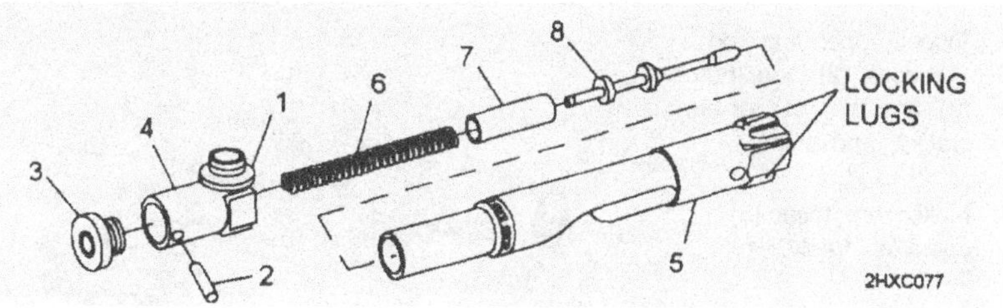

NOTE

Rotate cam actuator assembly so holes line up with headless straight pin. Use reamer of combination wrench to remove or replace headless straight pin.

3 Remove headless straight pin (2) and unscrew bolt plug assembly (3). Remove cam actuator assembly (4) from breech bolt (5). Separate spring (6), guide (7). and firing pin (8).

4 Make sure spring (6) is not kinked.

5 inspect threads on bolt plug assembly (3) and in breech bolt (5) for damage.

6 Clean powder fouling, corrosion, and dirt from all parts with CLP/RBC (items 1 and 2,WP 0039 00).

7 Drop firing pin (8) into breech bolt (5). Assemble and drop guide (7) and spring (6) into breech bolt (5).

8 Assemble cam actuator assembly (4) on breech bolt with roller to front. Screw in bolt plug assembly (3) and align three holes. Insert headless straight pin (2). Make sure cam actuator assembly (4) rotates freely when headless straight pin (2) is fully seated.

END **OF** TASK

MAINTENANCE OF OPERATING ROD ASSEMBLY 0020 00

INSPECT AND CLEAN

1 Inspect operating rod (1), yoke (2), sear notch (3), and pins (4) for burrs, cracks, and chips.

2 Make sure roller (5) operates freely.

3 Inspect driving spring (6) for tension, kinks, breaks, and wear.

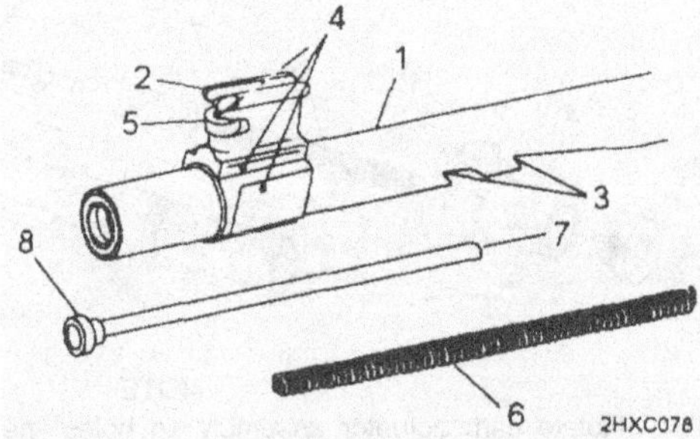

4 Check guide assembly (7) to make sure stop (8) is tight. If loose, notify unit maintenance.

5 Clean powder fouling, corrosion, and dirt from all parts with CLP/RBC (items land 2, WP 0039 00).

INSPECT AND CLEAN (cont)

6 Clean fluted carbon remover (9),

NOTE

Use fluted carbon remover (9) to clean inside of receiver (see WP 0025 00).

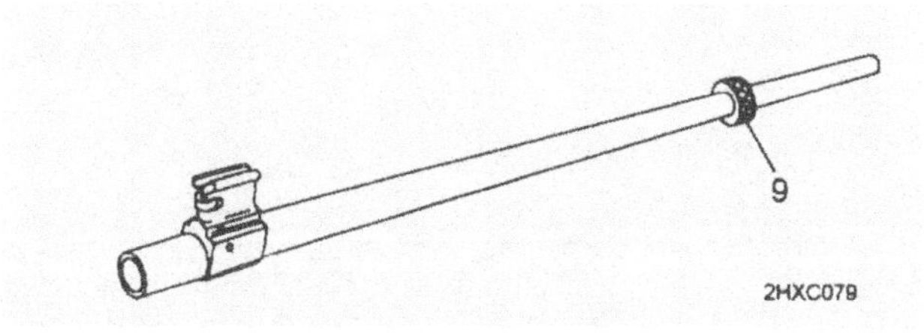

END OF TASK

MAINTENANCE OF COVER ASSEMBLY (M60 AND M60D) 0021 00

INSPECT AND CLEAN

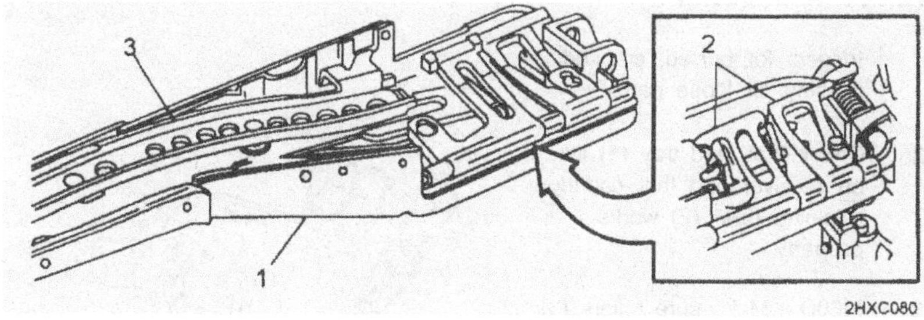

1 Inspect cover assembly (1) for bent, burred, cracked, loose, or missing parts.

2 Check spring action of front and rear cartridge guides (2), feed pawls (beneath the cartridge guides), and feed cam assembly (3).

INSPECT AND CLEAN (cont)

3 Make sure feed cam assembly is secure and operates freely.

4 Clean powder fouling, corrosion, and dirt from all areas with CLP/RBC (items 1 and 2, WP 0039 00).

END OF TASK

MAINTENANCE OF M60 HANGER AND CARTRIDGE FEED TRAY ASSEMBLY OR M60D CARTRIDGE FEED TRAY ASSEMBLY 0022 00

INSPECT AND CLEAN

1 Inspect for burred, cracked, missing, or loose parts.

2 Check that feed tray (1) fits on receiver and that cartridge retaining pawl (2) works properly.

3 M60D - Make sure rollers (3) do not bind.

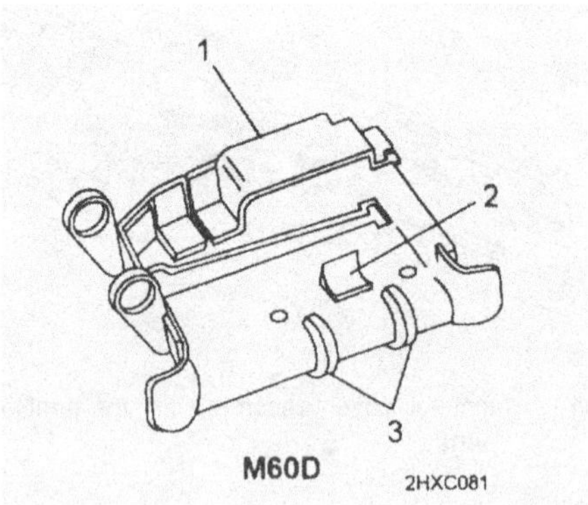

MAINTENANCE OF M60 HANGER AND CARTRIDGE FEED TRAY ASSEMBLY OR M60D CARTRIDGE FEED TRAY ASSEMBLY (cont) 0022 00

INSPECT AND CLEAN (cont)

4 M60 - Inspect for obstructions in hanger (4).

5 Clean corrosion and dirt from ail areas with CLP/RBC (items 1 and 2, WP 0039 00).

6 Remove CLP/RBC from top surface of feed tray (1) if gun is to be fired.

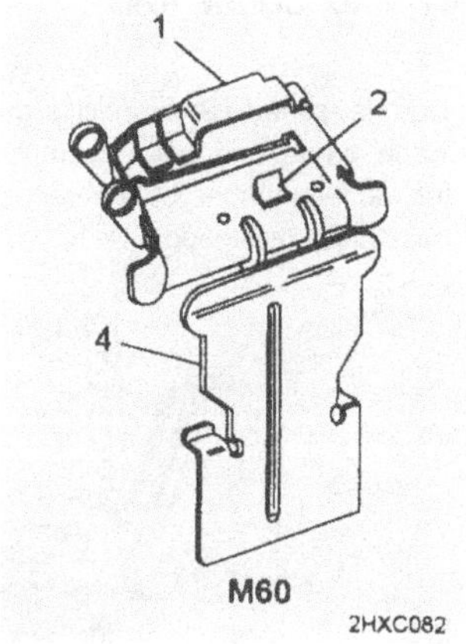

END OF TASK

MAINTENANCE OF BARREL AND BIPOD ASSEMBLY (INCLUDING GAS SYSTEM) 0023 00

BIPOD ASSEMBLY

INSPECT AND CLEAN

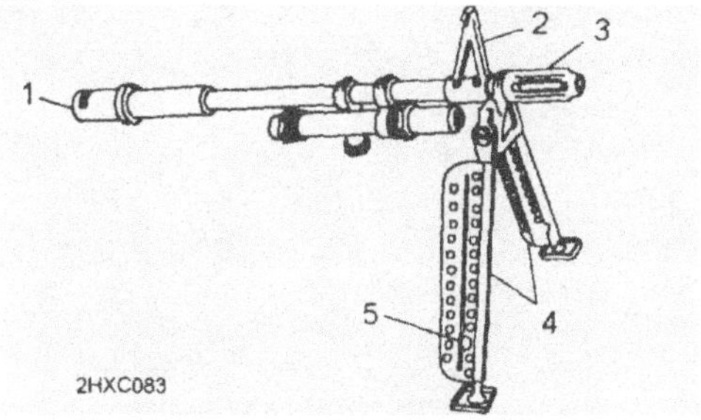

1 Check for burrs, cracks, and wear especially in the barrel socket area (1).

2 Make sure sight (2) and flash suppressor (3) are tight.

3 Make sure bipod (4) works, its legs are straight, and connections are tight.

4 Make sure bipod plungers (5) operate smoothly.

BIPOD ASSEMBLY (cont)

INSPECT AND CLEAN (cont)

WARNING

Barrels issued for a specific gun will not be changed from gun to gun. Each barrel and bolt assembly should be tagged during cleaning with the serial number of the receiver (gun) and should remain together as initially assigned.

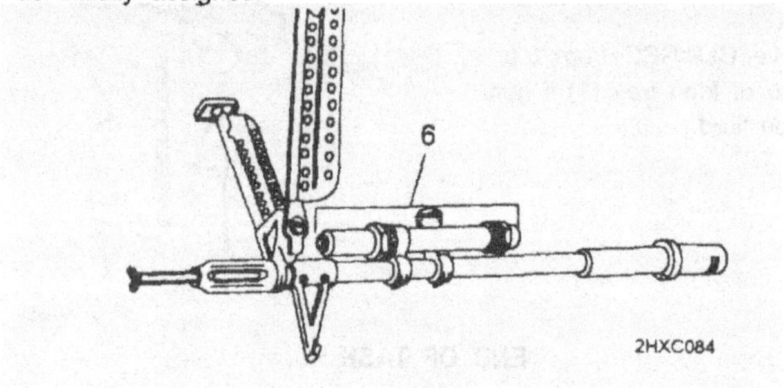

5 Turn barrel upside down for cleaning. This will keep gas piston and inside of gas cylinder (6) dry.

6 Run a cleaning rod with swab (item, 6 WP 0039 00) saturated with CLP/ RBC (items 1 and 2,WP 0039 00) through bore. Leave bore wet.

GAS SYSTEM

INSPECT AND CLEAN

CAUTION

No abrasive materials will be used to clean the gas piston or inside the gas clinder.

NOTE

Gas piston clicks when barrel is tilted. Take gas system (gas cylinder) apart only when you don't hear a click.

1 Use combination wrench and clean swab (item 6, WP 0039 00) to clean powder fouling, corrosion, and dirt from all parts.

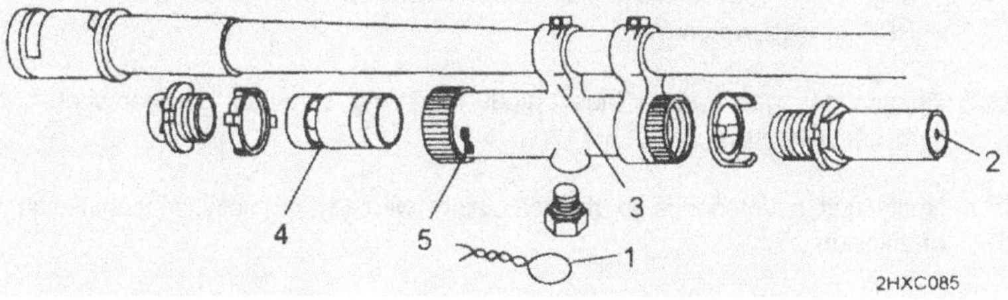

2 Remove safety wire (1) (notify unit maintenance).

3 Disassemble gas cylinder with combination wrench

4 Clean extension vent hole (2) with safety wire.

5 Clean gas port (3), gas piston hole (4), and gas cylinder vent holes (5) with reamer portion of wrench.

GAS SYSTEM (cont)

INSPECT AND CLEAN (cont)

CAUTION

Insert reamer carefully to prevent damage.

6 Bottom the reamer in the hole in the gas port (3) to make sure you have removed all carbon. Run small arms cleaning brush (bore) through bore to remove carbon deposits.

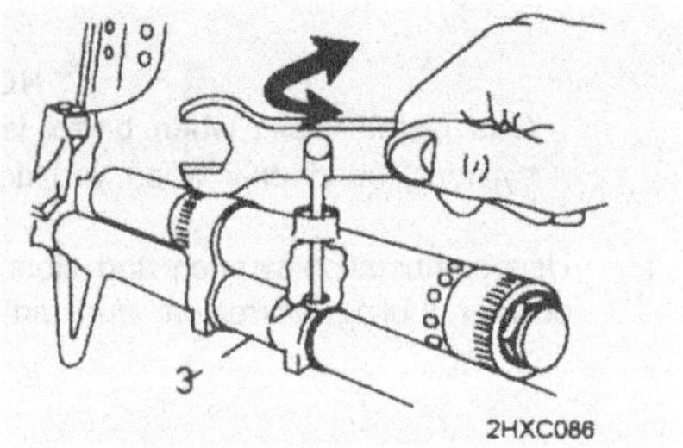

CAUTION

Make sure you do not install piston backward. Do not get CLP/RBC in gas system.

7 Reassemble gas system (gas cylinder), making sure gas cylinder vent, plug and nut are tight.

8 Notify unit maintenance to replace safety wire (1) promptly on completion of mission.

9 When safety wire is not available, check gas system periodically for tightness.

END OF TASK

INSPECT AND CLEAN

1 Hold down on sear (1) and remove sear pin (2).

2 Remove sear (11, sear plunger (3), and spring (4).

3 Remove trigger pin (5) and trigger (6).

4 Clean powder fouling, corrosion, and dirt from all areas with CLP/RBC (items 1 and 2, WP 0039 00) and wipe with a small arms cleaning brush.

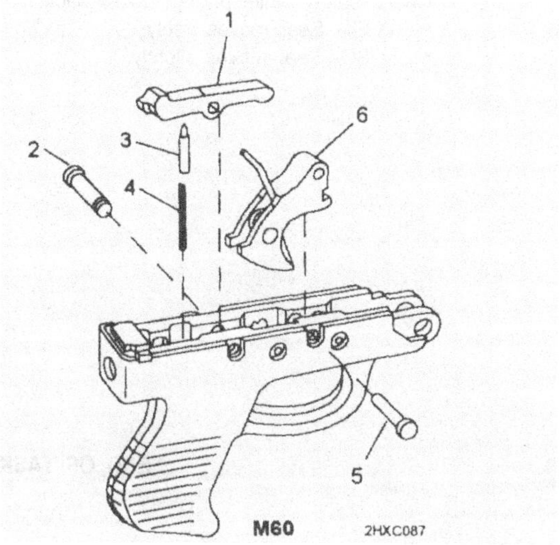

INSPECT AND CLEAN (cont)

5 Inspect sear (1) for chips, cracks and signs of wear.

6 Inspect sear plunger (3) and spring (4) for signs of wear.

7 Position trigger (6) in the grip so that the trigger spring is under the channel. Align holes and insert trigger pin (5) through the right side of grip and secure trigger (6).

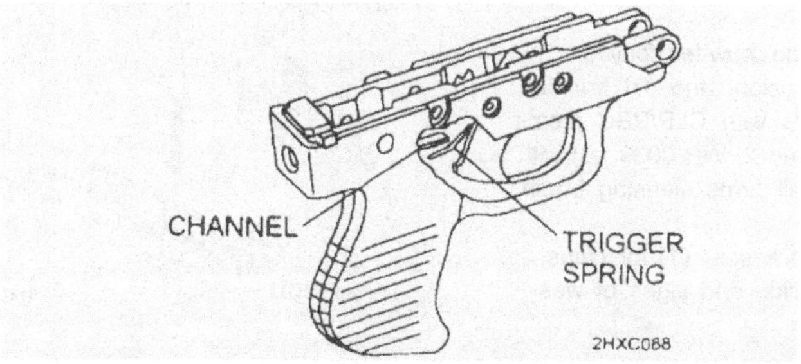

8 Insert spring (4) and sear plunger (3) into hole. Position sear (1) and apply slight pressure to align holes. Insert sear pin (2) through left side of grip and secure sear.

9 Check safety while pulling trigger back:
- 'F' - Sear moves freely.
- 'S' - Sear moves slightly.

END OF TASK

MAINTENANCE OF M60D SEAR AND SAFETY HOUSING ASSEMBLY 0025 00

INSPECT AND CLEAN

1 Hold down top of sear (1) and remove sear pin (2) sear assembly activator (3) and sleeve bushing (4).

2 Lift out sear (1) sear plunger (5), and spring (6).

3 Clean powder fouling, corrosion, and dirt from all parts with CLP/RBC (items 1 and 2, WP 0039 00) with a small arms cleaning brush.

4 Check sear (1) for chips, cracks, and signs of wear.

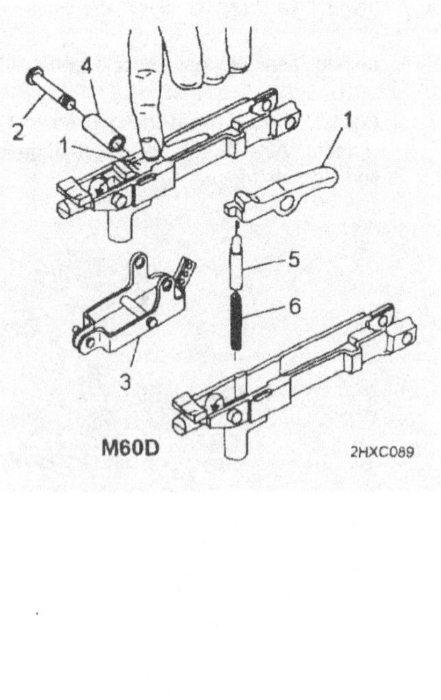

INSPECT AND CLEAN (cont)

5 Check sear plunger (5) and spring (6) for signs of wear.

6 Insert spring (6) and sear plunger (5) into hole. Position sear (1) into housing.

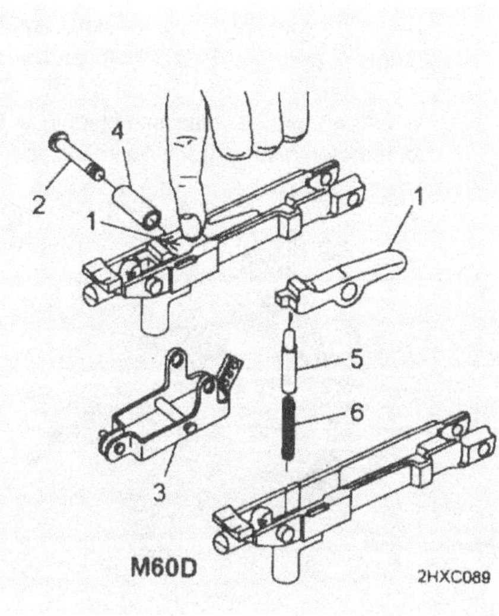

7 Apply slight pressure to align holes and insert sleeve bushing (4).

8 Position sear assembly activator (3) on bottom of sear housing and secure with sear pin (2).

9 Check safety while pulling down on sear assembly activator.

- 'F' - Sear moves freely.
- 'S' - Sear moves slightly.

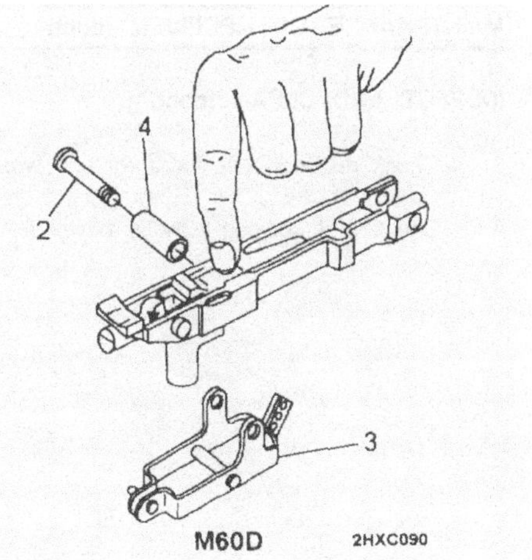

END OF TASK

INSPECT AND CLEAN

1 Clean powder fouling, corrosion, and dirt from all areas with CLP/RBC (items 1 and 2, WP 0039 00) and small arms brushes.

2 When weapon is disassembled, use fluted end of operating rod assembly to remove carbon from operating rod guide of receiver (WP 0020 00).

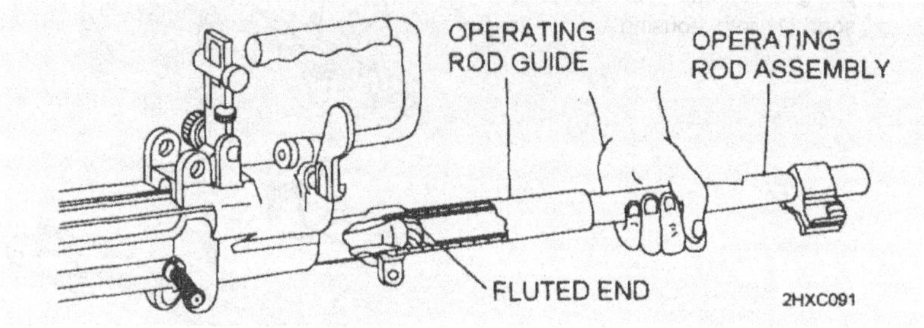

INSPECT AND CLEAN (cont)

3 Check receiver rails for burrs and wear.

4 Check that cocking handle moves freely.

END OF TASK

M60 FOREARM ASSEMBLY

CAUTION

When installing forearm assembly, be careful not to damage its internal ribs.

1 Slightly tip forearm assembly and slide it onto receiver.

2 Press in on bottom of forearm assembly to latch.

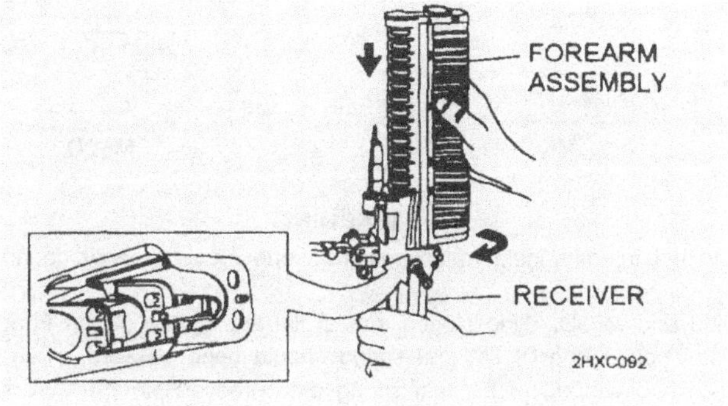

M60 TRIGGER MECHANISM GRIP ASSEMBLY OR M60D SEAR AND SAFETY HOUSING ASSEMBLY

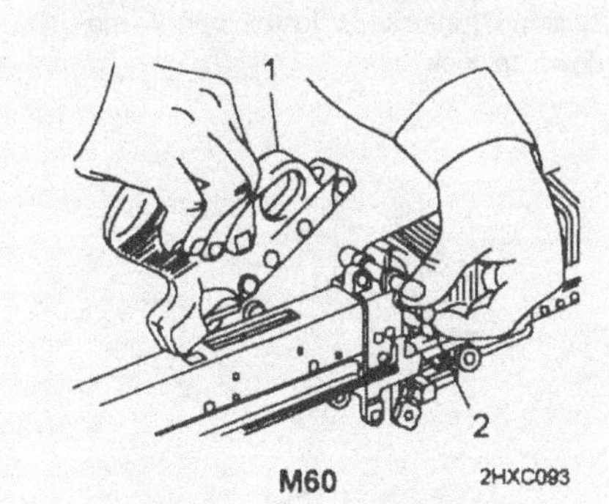

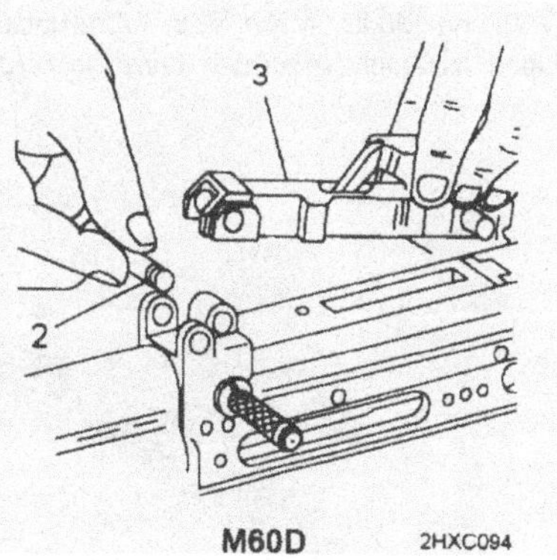

1 M60. Position trigger mechanism assembly (1) on bottom of receiver, align holes, and install front pin (2) from the left side.

M60D. Position sear and safety housing assembly (3) on bottom of receiver, align holes and install front pin (2) from the left side.

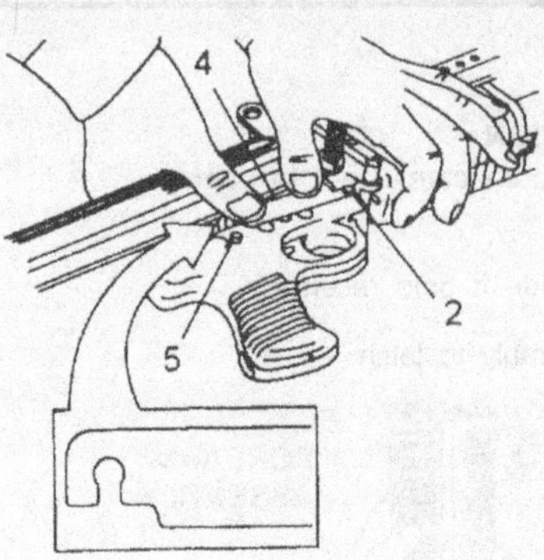

M60 2HXC095

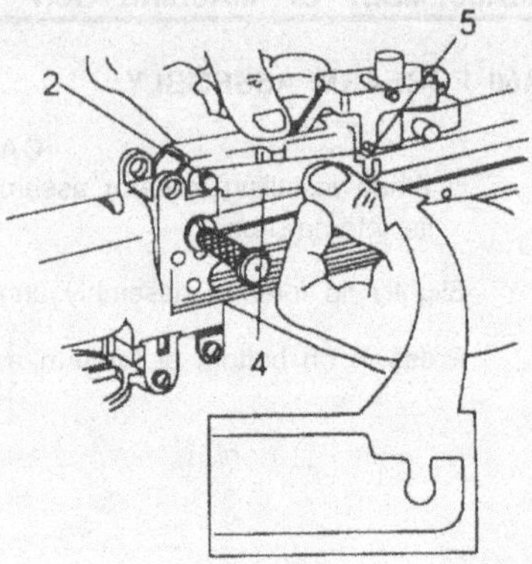

M60D 2HXC096

WARNING

Do not interchange leaf springs from one model weapon to another.

2 M60 and M60D. Slide slotted end of flat leaf spring (4) on front pin (2). The hooked end of flat leaf spring should bend outward. Push down and slide hooked end of flat leaf spring onto grooved pin (5).

REASSEMBLY OF MACHINE GUN (cont) 0027 00

BARREL AND BIPOD ASSEMBLY

M60 and M60D. Make sure barrel lock (3) is up (unlocked). Install barrel and bipod assembly (4). Push barrel lock (3) down to lock.

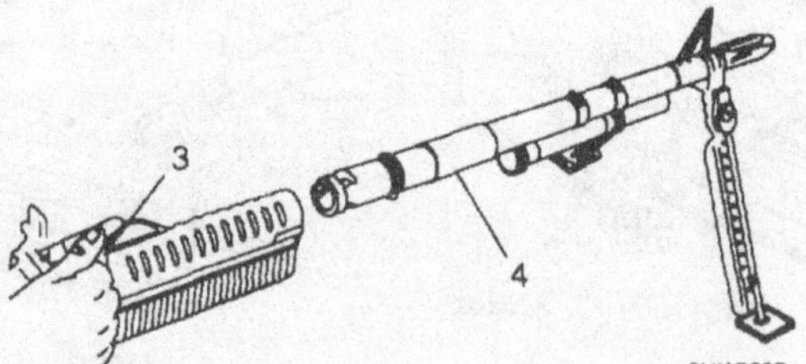

2HXC097

WARNING

To prevent possible bodily injury and aircraft damage, personnel should not stand the weapon on its barrel assembly when disassembling or assembling the weapon.

M60 D. Raise gun adapter (1), align holes, and insert quick release pin (2) to secure gun adapter to receiver.

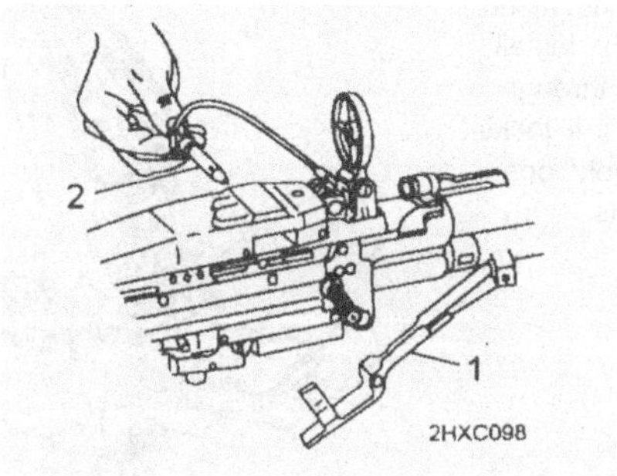

COVER ASSEMBLY AND M60 HANGER AND CARTRIDGE FEED TRAY ASSEMBLY OR M60D CARTRIDGE FEED TRAY ASSEMBLY

1 M60 or M60D. Position M60 hanger and cartridge feed tray assembly (1) or M60D cartridge tray assembly (2) on receiver.

2 Install torsion spring (3) and cover assembly (4). Make sure ends of torsion spring stick in holes of cover and receiver.

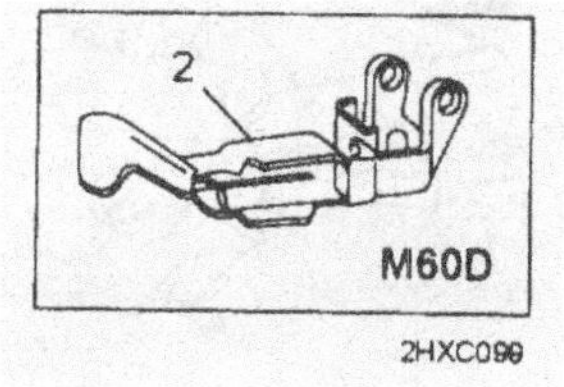

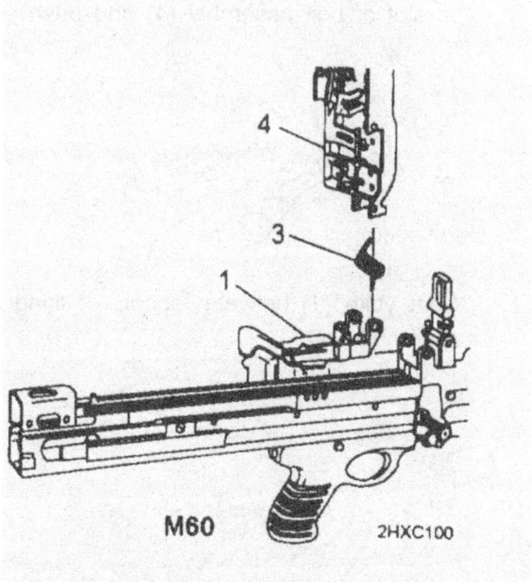

NOTE

Insert the hinge pin from one side and the hinge pin latch from the opposite side.

3 Apply slight pressure to line up torsion spring (3) and insert cover hinge pin (5). Install hinge pin latch (6) through cover hinge pin until it locks. Keep cover assembly open until machine gun is reassembled.

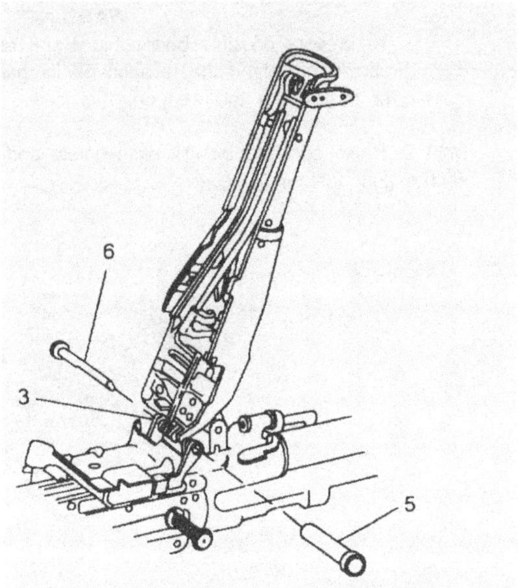

2HXC101

BOLT ASSEMBLY AND OPERATING ROD ASSEMBLY

1 Place yoke (1) of operating rod assembly (2) against spool of firing pin (3) in slot of bolt assembly (4) and push spool in direction of arrow.

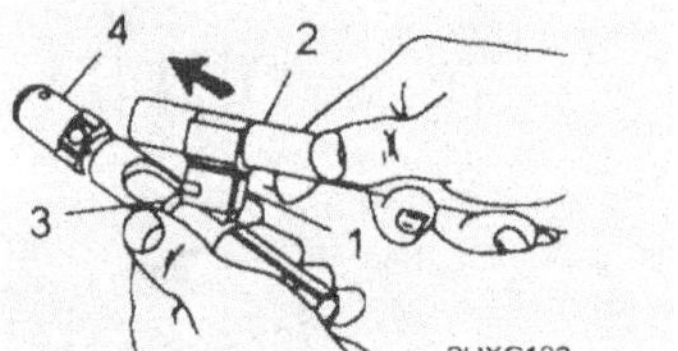

2HXC102

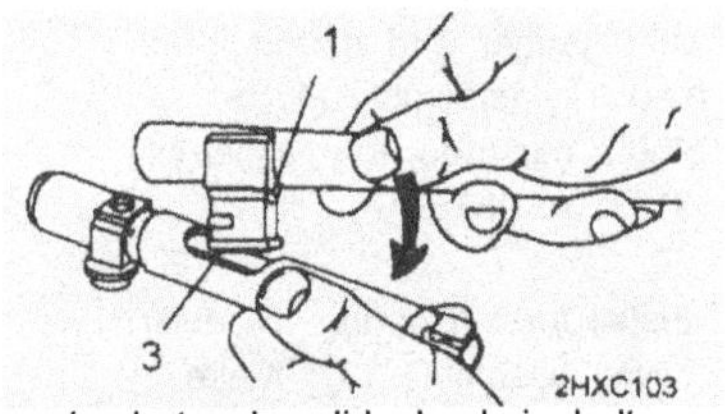

2HXC103

2 Seat yoke (1) between spools of firing pin. Let yoke slide back in bolt assembly (4).

2HXC104

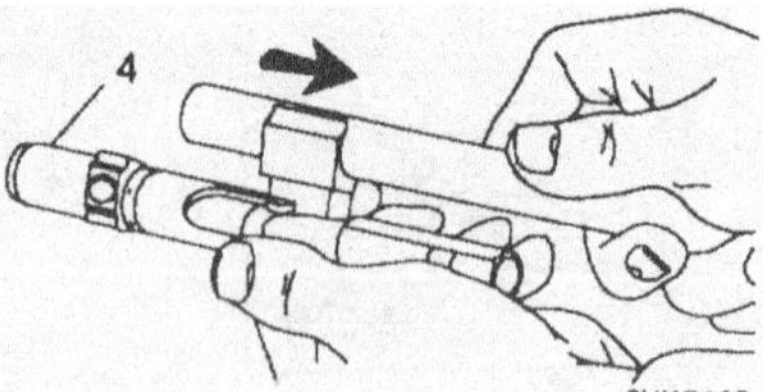

2HXC105

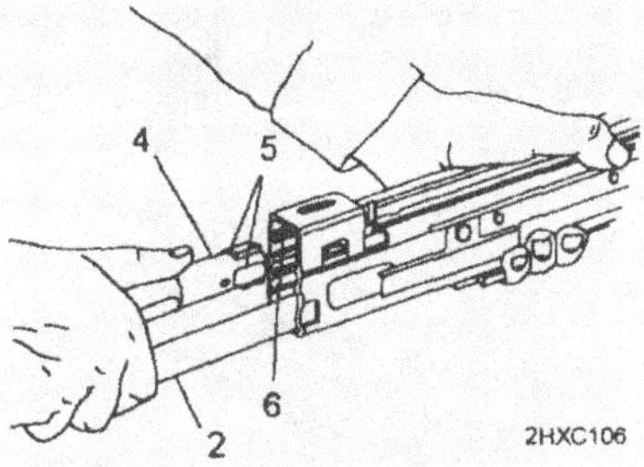

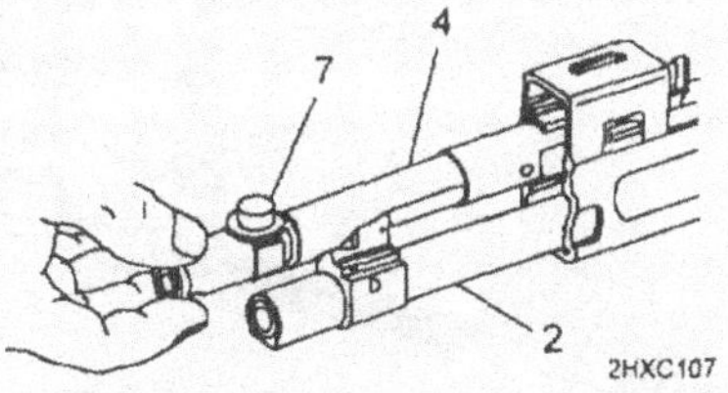

3 Start operating rod assembly (2) with bolt assembly (4) into weapon. Turn bolt assembly (4) so that stripping lugs (5) line up with upper rails (6). Push in until stripping lugs (5) engage rails (6).

4 Turn roller (7) straight up and push bolt assembly (4) and operating rod assembly (2) as a unit into weapon.

BOLT ASSEMBLY AND OPERATING ROD ASSEMBLY (cont)

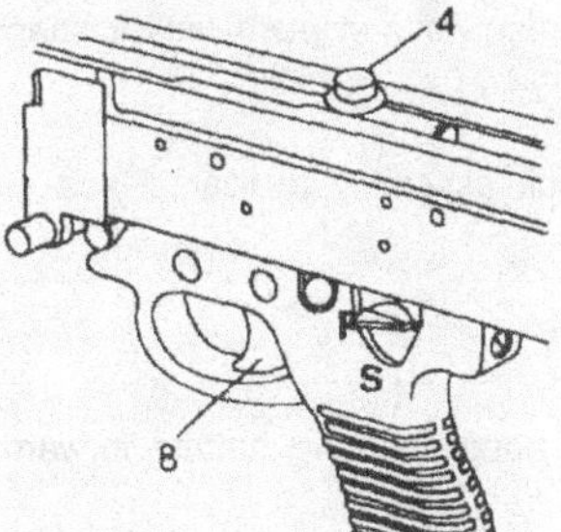

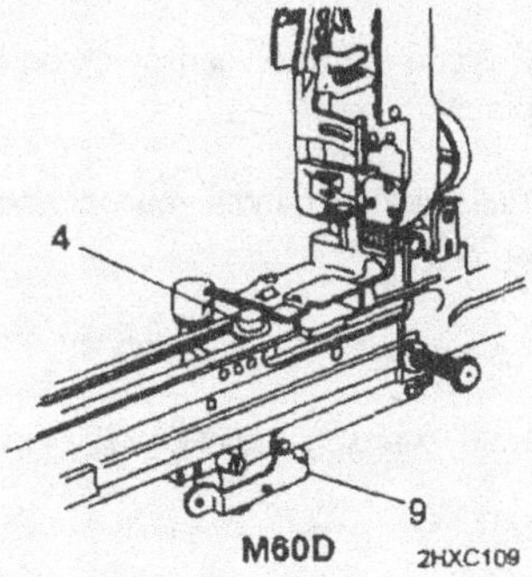

5 M60. Place safety to 'F'. Pull back on trigger (8) and hold while pushing bolt assembly (4) into receiver until it locks in place.

M60 D. Press in on 'F'. Pull down and hold sear assembly activator (9) while pushing bolt assembly (4) into receiver until it locks in place.

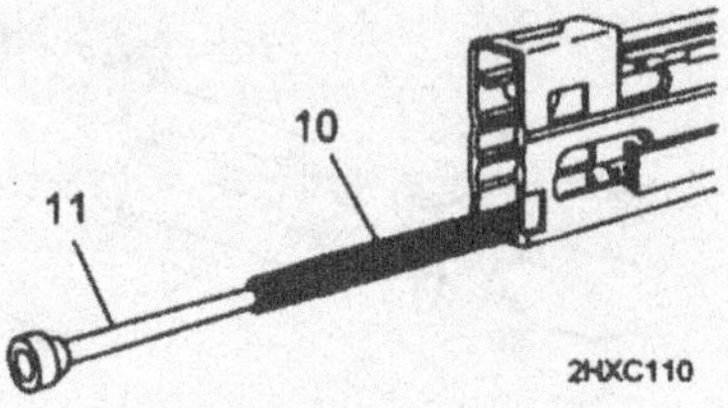

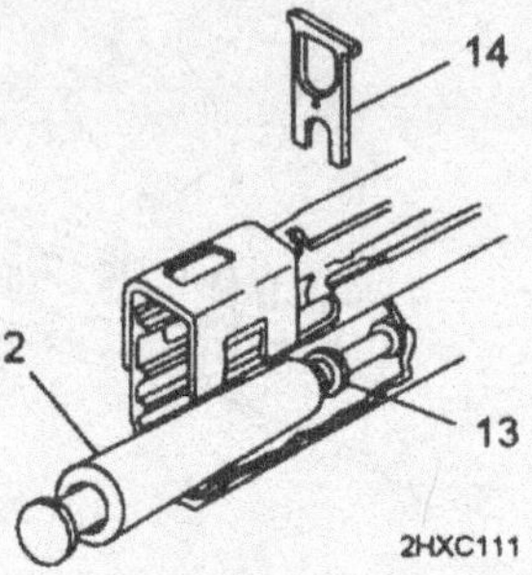

6 Install drive spring (10) and guide assembly (11) in rear of receiver.

7 Insert buffer assembly (12) against end of guide assembly and push until groove (13) on buffer assembly lines up with slot for buffer retaining yoke (14).

6 Insert buffer retaining yoke(14).

M60 SHOULDER GUN STOCK

1 Position shoulder gun stock on rear of receiver and push until it snaps in place.

2 Pull cocking handle rearward to lock bolt assembly to rear. Place safety to 'S'.

3 Close cover, move safety to 'F' position.

4 Hold charging handle, pull trigger, and ease charging handle forward.

M60 SPADE AND TRIGGER GRIP ASSEMBLY

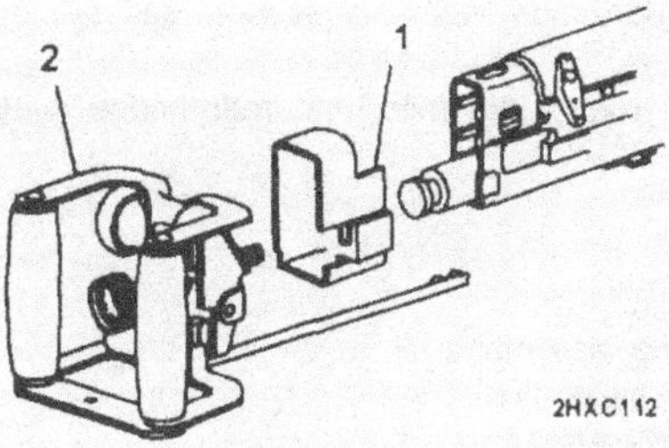

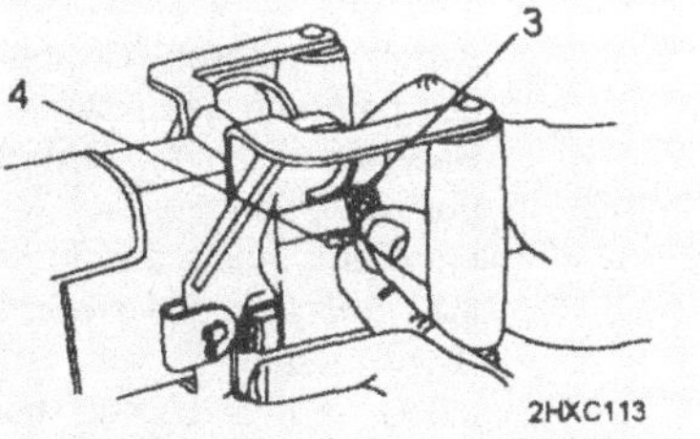

1 Install dust and moisture boot (1) and position grip and trigger assembly (2) on receiver.

2 Turn knob and pin assembly (3) until clicking is heard. Then press and hold spring pin (4) and turn knob (3) until tight.

3 Release spring pin (4) and turn knob and pin assembly (3) until it locks.

REASSEMBLY OF MACHINE GUN (cont) **0027 00**

M60 SPADE AND TRIGGER GRIP ASSEMBLY (cont)

4 Pull cocking handle rearward to lock bolt assembly to rear. Press in on 'S'.

5 Close cover (5).

6 Snap end of sear assembly link and spring (6) on end of sear assembly activator (7).

7 Press safety in on 'F', hold cocking handle, pull trigger, and ease **bolt** forward.

END OF TASK

MAINTENANCE CHECKLIST 0028 00

Check the cleanliness and overall condition of the following:

- Barrel assemblies. Chamber bore and flash suppressor, gas cylinder, gas port, and gas piston.
- Receiver assembly. Operating rod guide, rails, and rear bridge rivets.
- Bolt assembly.
- Front and rear sights.
- M60 Shoulder gun stock.
- Bipod assembly.
- M60D Sear and safety housing assembly.
- M60 Trigger mechanism grip assembly.
- M60D Spade and trigger grip assembly.
- Cocking handle assembly.
- Cover assembly and M60 hanger and cartridge feed tray assembly.
- Cover assembly and M60D cartridge feed tray assembly.
- Forearm assembly.
- Blank firing attachment (BFA). Clean after firing 500 rounds or after each day's firing, whichever comes first. Remember BFA causes carbon fouling.
- Small arms sling.

FUNCTION CHECK 0029 00

Function check shall be performed after any maintenance or disassembly of the weapon.

Bolt assembly rearward? Good. Safety in 'S'? Good. Gun cleared? Good.

1 Load machine gun with dummy ammunition (WP 0005 00).

2 Pull trigger. Nothing should happen.

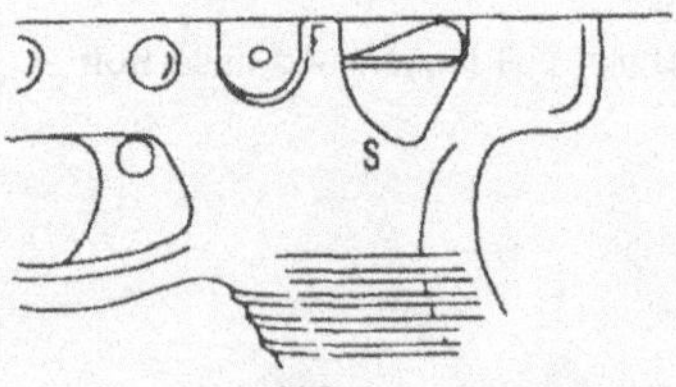

M60 2HXC114

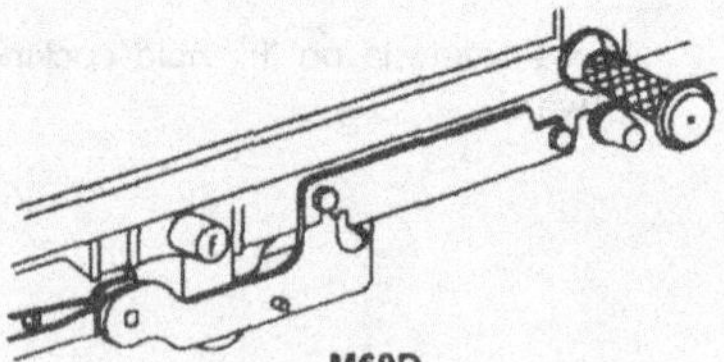

M60D 2HXC115

3 M60. Place safety to 'F'.
M60D. Press in on 'F'.

4 Pull trigger. Bolt assembly should move forward, strip and chamber a dummy round, and lock.

5 Pull cocking handle back. Dummy round should eject, and bolt assembly should lock to rear.

WARNING

After unloading dummy round, be sure the barrel is cleared.

6 M60. Place safety to 'S'.
M60D. Press in on 'S'.

7 Unload dummy ammunition and clear weapon (see WP 0009 00).

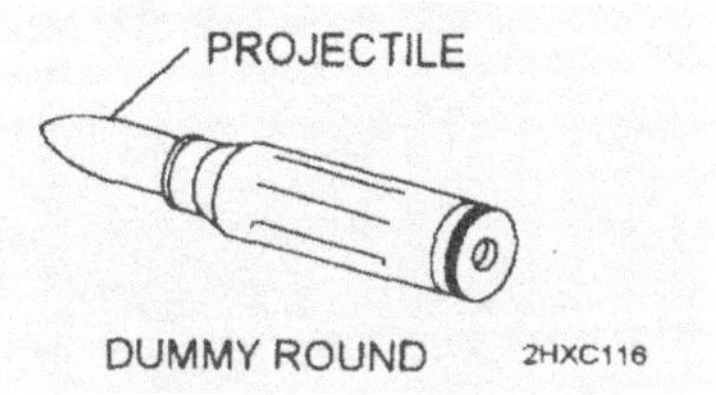

8 Be sure that projectile stays on dummy round.

9 Close **cover** and latch.

10 M60. Place safety to 'F'. Pull cocking handle, pull trigger, and ease bolt assembly forward.
M60D. Press in on 'F'. Pull cocking handle, pull trigger, and ease bolt assembly forward.

END OF TASK

CHAPTER 4

MAINTENANCE OF AUXILIARY EQUIPMENT

MAINTENANCE OF AUXILIARY EQUIPMENT 0030 00

MACHINE GUN M122 TRIPOD MOUNT. Refer to TM 9-1005-245-13&P.

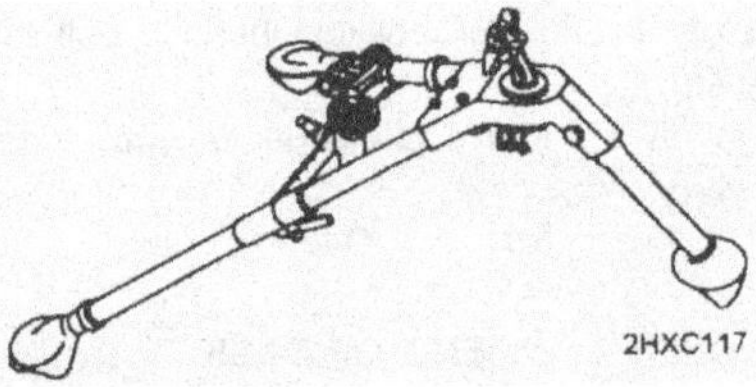

NIGHT VISION SIGHT AN/PVS-4. Refer to TM 11-5855-213-10 for maintenance.

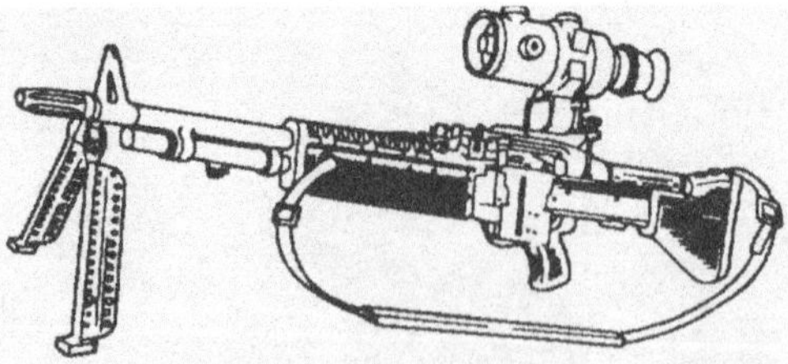

NIGHT VISION SIGHT

MOUNT PEDESTAL, MACHINE GUN, M6 AND MOUNT, MACHINE GUN M197. Refer to TM 9-1005-245-13&P

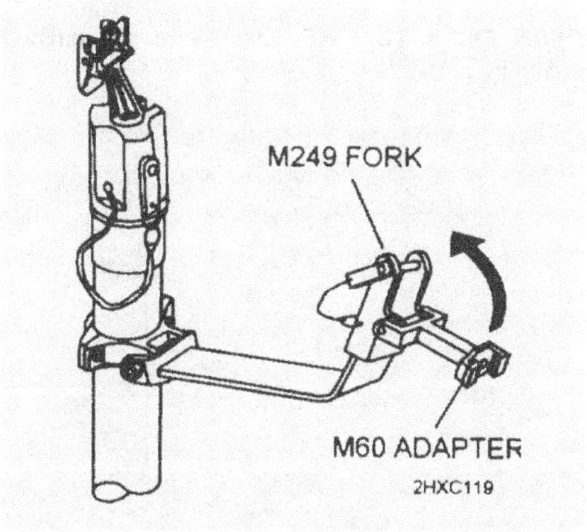

CHAPTER 5

AMMUNITION

AMMUNITION 0031 00

WARNING

This is the only ammunition authorized for use in your machine gun. If it s not shown, it is not authorized.

AMMUNITION WHICH FAILS TO FIRE. Dispose of any ammunition, which fails to fire, according to authorized procedures.

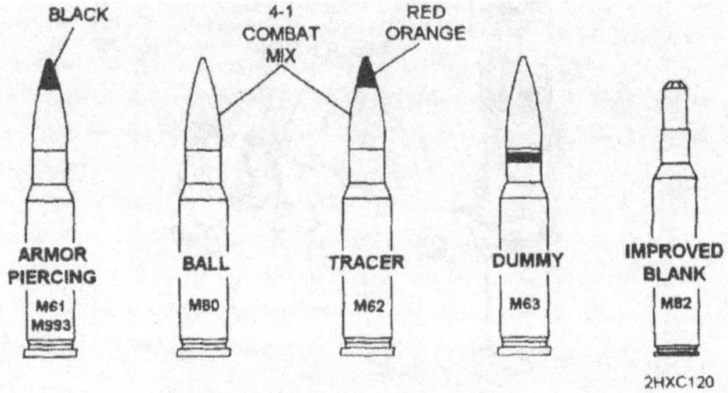

7.62-MM NATO AMMUNITION

AMMUNITION (cont) 0031 00

CARE, HANDLING, AND PRESERVATION.

a. Do not open ammunition containers until the ammunition is to be used. Ammunition removed from the airtight containers, particularly in damp climates, is likely to corrode.

b. Protect ammunition from mud, dirt, and water. If the ammunition gets wet or dirty, wipe it off prior to use. Wipe off light corrosion as soon as it is discovered. Heavily corroded cartridges or cartridges which have dented cases or loose projectiles should not be fired.

c. Do not expose the ammunition to the direct rays of the sun. If the powder is hot, excessive pressure may be developed when the gun is fired.

d. Do not oil or grease ammunition. Dust and other abrasives collecting on oiled or greased ammunition will damage the operating parts of the gun, and oil on cartridges will produce excessive chamber pressure.

REFERENCES

REFERENCES	0032 00

SCOPE

This appendix lists all forms, field manuals, and technical manuals referenced in this manual.

CONSOLIDATED TABLE OF ALLOWANCES

CTA 8-100 .... Army Medical Department Expendable/ Durable Items

CTA 50-970 .... Expendable/Durable Items (Except: Medical, Class V, Repair Parts, and Heraldic Items)

FIELD MANUALS

FM 3-87 ....,...................... Nuclear, Biological and Chemical (NBC) Reconnaissance and Decontamination Operations (How to Fight)

FM 21-11 .... First Aid for Sholdiers

FM 21-40 .... NBC (Nuclear, Biological and Chemical) Defense

FM 23-67 .... Machine-gun 7.62-MM, M60

FIRING TABLE

FT 7.62-A-2 .... Machine Gun, 7.62-MM: M60 on Mount, Machine Gun: 7.62-MM, M122 and Machine Gun, 7.62-MM: M73 on Tank Combat, Full-Tracked: 105-MM Gun, M60 Series and Rifle, 7.62-MM: M14: Firing Cartridge, 7.62-MM, Ball, NATO, M59; Cartridge, 7.62-MM, Ball, NATO, M80; Cartridge, 7.62-MM: AP, NATO, M61/M993 and Cartridge, 7.62-MM; Tracer, NATO, M62

FORMS

DA Form 2028 .... Recommended Changes to Publications and Blank Forms

SF 368 .... Quality Deficiency Report

TECHNICAL MANUALS

TM 3-220 .... Chemical, Biological, and Radiological (CBR) Decontamination

TM 11-5855-213-10 .... Operator's Manual for Night Vision Sight, Individual Served Weapon, AN/PVS-4

TM 9-1005-245-13&P .... Machine Gun Mounts and Combinations for Tactical/Armored Vehicles and Ground Mounting

TM 9-4933-273-12&P .... Boresighting Equipment, Weapon Small Arms, M30

MISCELLANEOUS PUBLICATIONS

DA PAM 738-750 The Army Maintenance Management System (TAMMS)

COMPONENTS OF END ITEM

AND

BASIC ISSUE ITEMS LISTS

INTRODUCTION 0033 00

SCOPE

These work packages lists COEI and BII for the M60/M60D machine gun to help you inventory items required for safe and efficient operation of the equipment.

GENERAL

The Components of End Item and Basic Issue Items Lists are divided into the following lists:

Components of End Item (COEI). This list is for informational purposes only and is not authority to requisition replacements. These items are part of the M60/M60D machine gun. As part of the end item, these items must be with the end item whenever it is issued or transferred between property accounts. Items of COEI are removed and separately packaged for transportation or shipment only when necessary. Illustrations are furnished to help you find and identify the items.

GENERAL (cont)

Basic Issue Items (BII). These essential items required to place the M60/M60D machine gun in operation, operate it, and to do emergency repairs. Although shipped separately packaged, BII must be with the M60/M60D machine gun during operation and whenever it is transferred between property accounts. Listing these items is your authority to request/requisition them for replacement based on authorization of the end item by the TOE/MTOE. Illustrations are furnished to help you find and identify the items.

EXPLANATION OF COLUMNS IN THE COEI LIST AND BII LIST

Column (1) - Illus Number. Gives you the number of the item illustrated.

Column (2) - National Stock Number (NSN). Identifies the stock number of the item to be used for requisitioning purposes.

Column (3) - Description, CAGEC, and Part Number. Identifies the Federal name (in all capitol letters) followed by a minimum description when needed. The stowage location of COEI and BII is also included in the column. The last line below the description is the CAGEC (Commercial and Government Entity Code) (in parentheses) and the part number.

Column (4) - Usable On Code. When applicable, gives you a code if the item you need is not the same for different models of equipment.

Code	Used On
023	Model M60
G79	Model M60D

Column (5) - Unit of Measure (U/M). Indicates the physical measurement or count of the item as issued per the National Stock Number shown in column (2).

Column (6) - Qty Rqr. Indicates the quantity required.

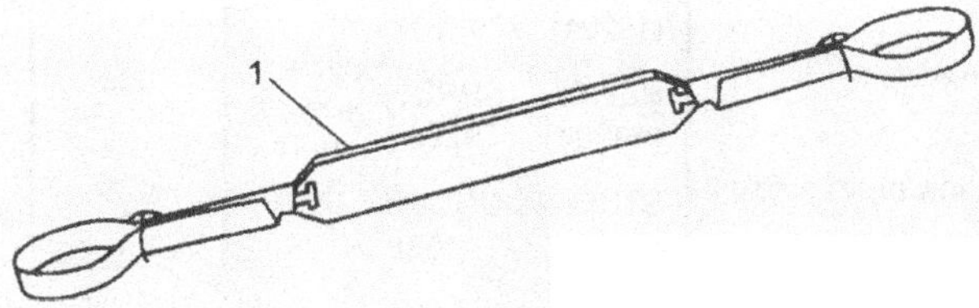

(1) ILLUS NO.	(2) NATIONAL STOCK NUMBER	(3) DESCRIPTION, CAGEC. AND PART NUMBER	(4) USABLE ON CODE	(5) U/M	(6) QTY RQR
1	1002-00-312-7177	SLING, SMALL ARMS (19204) 12003483	023	EA	1

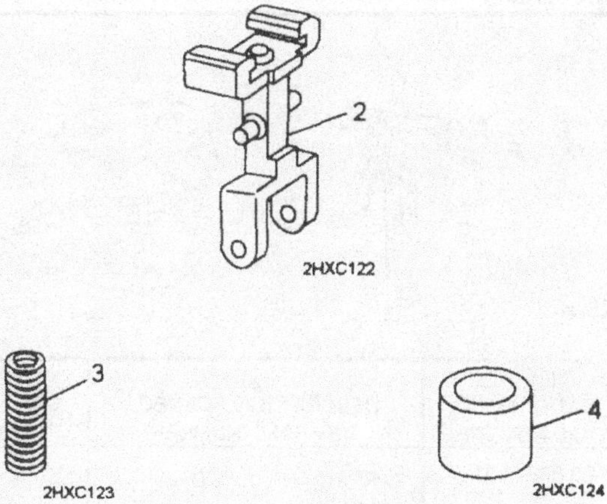

(1) ILLUS NO.	(2) NATIONAL STOCK NUMBER	(3) DESCRIPTION, CAGEC, AND PART NUMBER	(4) USABLE ON CODE	(5) U/M	(6) QTY RQR
2	1005-00-772-0194	ADAPTER ASSEMBLY (19204) 7792991	023	EA	1
3	5360-00-772-0193	HELICAL SPRING (19204) 7792987	023	EA	1
4	1005-00-772-0192	STUD, PLUNGER (19204) 7792986	023	EA	1

BASIC ISSUE ITEMS **0035 00**

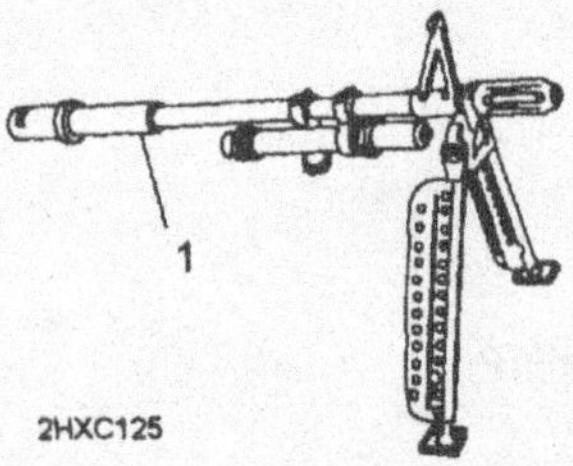

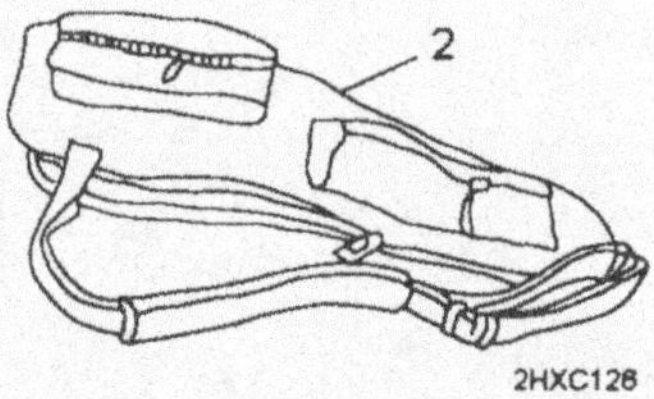

(1) ILLUS NO.	(2) NATIONAL STOCK NUMBER	(3) DESCRIPTION, CAGEC, AND PART NUMBER	(4) USABLE ON CODE	(5) U/M	(6) QTY RQR
1	1005-00-608-5001	BARREL AND BIPOD ASSEMBLY (19204) 7269027		EA	1
2	1005-00-791-5420	CASE, CARRYING, BARREL ASSEMBLY AND EQUIPMENT (19205) 7791009		EA	1

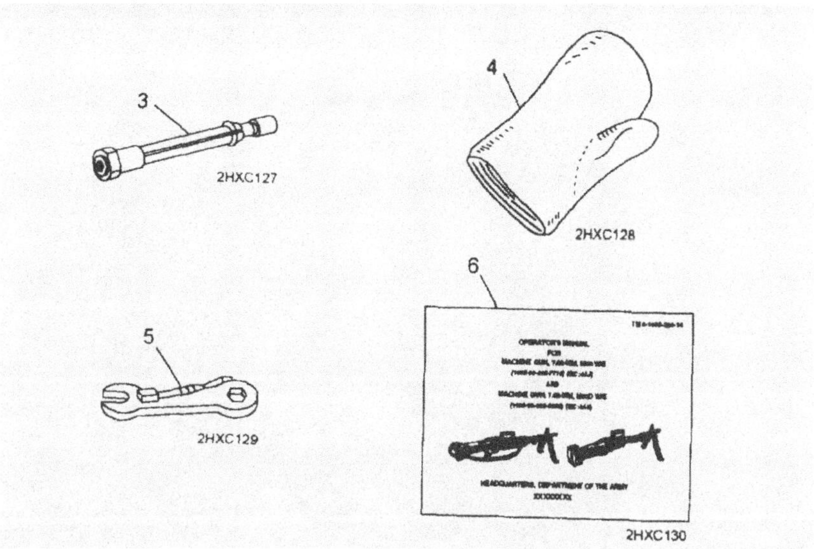

(1) ILLUS NO.	(2) NATIONAL STOCK NUMBER	(3) DESCRIPTION, CAGEC. AND PART NUMBER	(4) USABLE ON CODE	(5) U/M	(6) QTY RQR
3	4933-00-652-9950	EXTRACTOR, RUPTURED CARTRIDGE CASE (19205) 7790352		EA	1
4	8415-01-092-0039	MITTEN, HEAT PROTECTIVE (81349) MIL-M-11199		EA	1
5	5120-00-461-1075	WRENCH, SCREWDRIVER AND REAMER (COMBINATION) (19204) 8448458		EA	1
6		TM 9-1005-224-10		EA	1

ADDITIONAL AUTHORIZED LISTS

INTRODUCTION 0036 00

SCOPE

This work package lists additional items you are authorized for the support of the M60/M60D machine Gun.

GENERAL

This list identifies items that do not have to accompany the M60/M60D machine gun and that do not have to be turned in with it. These items are all authorized to you by CTA, MTOE, TDA, or JTA.

EXPLANATION OF COLUMNS IN AAL

Column (1) - National Stock Number (NSN). Identifies the stock number of the item to be used for requisitioning purposes.

Column (2) - Description, Commercial and Government Entity Code (CAGE), and Part Number (P/N). identifies the Federal item name (in all capital letters) followed by a minimum description when needed. The last line below the description is the CAGEC (in parentheses) and the part number.

EXPLANATION OF COLUMNS IN AAL (cont)

Column (3) - Usable On Code. When applicable, gives you a code if the item you need is not the same for different models of equipment.

Code	Used On
023	Model M60
G79	Model M60D

Column (4) - Unit of Measure (U/M). Indicates the physical measurement or count of the item as issued per the National Stock Number shown in column (1).

Column (5) - Qty Recm. Indicates the quantity recommended.

ADDITIONAL AUTHORIZED LIST 0037 00

(1) NATIONAL STOCK NUMBER	(2) DESCRIPTION, CAGEC. AND PART NUMBER	(3) USABLE ON CODE	(4) U/M	(5) QTY RECM
	MTOE AUTHORIZED ITEMS			
5855-00-629-5334	SIGHT, NIGHT VISION AN/PVS-4 (60063) SMD850300-1	023	EA	1
1005-00-710-5599	M122 MACHINE GUN TRIPOD (19205) 7790723		EA	1
	CTA AUTHORIZED ITEMS			
1005-00-556-4174	BRUSH, CLEANING, SMALL ARMS: (BORE) (19204) 5564174		EA	2
1005-00-690-3115	BRUSH, CLEANING, SMALL ARMS: (CHAMBER) (19204) 7790452		EA	1
1005-00-350-4100	BRUSH, CLEANING, SMALL ARMS: (RECEIVER) (19204) 8848466		EA	1

(1) NATIONAL STOCK NUMBER	(2) DESCRIPTION, CAQEC, AND PART NUMBER	(3) USABLE ON CODE	(4) U/M	(5) QTY RECM
	CTA AUTHORIZED ITEMS (cont)			
1005-00-140-3515	FIRING ATTACHMENT, BLANK (19204) 12002981		EA	1
1005-00-793-6761	HANDLE ASSEMBLY: CLEANING ROD (19204) 7266115		EA	1
1005-00-726-6109	ROD, SECTIONS, CLEANING, SMALL ARMS (19204) 7266109		EA	5
	SWAB HOLDER SECTION, SMALL ARMS CLEANING ROD (19204) 7266110		EA	1

EXPENDABLE/DURABLE SUPPLIES AND MATERIALS LIST

SCOPE

This work package lists expendable and durable items that you will need to operate and maintain the M60/M60D machine gun. This list is for information only and is not authority to requisition the listed items. These items are authorized to you by CTA 50-970, Expendable/Durable Items (Except Medical, Class V Repair Parts, and Heraldic Items), or CTA 8-100, Army Medical Department Expendable/Durable Items.

EXPLANATION OF COLUMNS IN THE EXPENDABLE/DURABLE ITEMS LIST

Column (1) - Item Number. This number is assigned to the entry in the list and is referenced in the narrative instructions to identify the item (e.g., Use wiping rag, (item 5, WP 0034 00).)

Column (2) - Level, This column identifies the lowest level ofmaintenance that requires the listed item.

C - Operator/Crew

EXPLANATION OF COLUMNS IN THE EXPENDABLE/DURABLE ITEMS LIST (cont)

Column (3) - National Stock Number (NSN). This is the NSN assigned to the item which you can use to requisition it.

Column (4) - Item Name, Descprition, Commercial and Government Entity Code (CAGEC), and Part Number (P/N). This column provides the other information you need to identify the item.

Column (5) - Unit of Measure (U/M). This code shows the physical measurement or count of an item, such as gallon, dozen, gross, etc.

EXPENDABLE/DURABLE SUPPLIES AND MATERIALS LIST

0039 00

(1) ITEM NUMBER	(2) LEVEL	(3) NATIONAL STOCK NUMBER	(4) DESCRIPTION, CAGEC, AND PART NUMBER	(5) U/M
1	C		Cleaner, Lubricant and Preservative: grade 2 (CLP)	
		9150-01-102-1473	1/2 oz bottle	EA
		9150-01-079-6124	4 oz bottle	EA
			(8 1349) MIL-L-63460	
2	C		Cleaning Compound, Solvent: Rifle Bore Cleaner (RBC) (81349) MIL-C- 372	
		6850-00-224-6656	2 oz can	EA
		6850-00-224-6657	8 oz can	EA
3	C		Lubricating Oil, Weapons (LAW) (81349) MIL-L-14107	
		9150-00-292-9689	1 qt can	EA

EXPENDABLE/DURABLE SUPPLIES AND MATERIALS LIST (cont)

0039 00

(1) ITEM NUMBER	(2) LEVEL	(3) NATIONAL STOCK NUMBER	(4) DESCRIPTION, CAGEC, AND PART NUMBER	(5) U/M
4	C		Lubricating Oil, Weapons: semi-fluid (LSA) (19204) 8436793	
		9150-00-889-3522	4 oz bottle	EA
5	C		Rag, Wiping: (cotton) (58536) A-A-531	
		7920-00-205-1711	50 lb bdl	LB
6	C		Swab, Small Arms Cleaning: (cotton) (19204) 5019316	
		1005-00-288-3565	(1000 per pkg)	PG

Subject **WP Sequence No.**

Subject **WP Sequence No.**

Subject **WP Sequence No.**

Subject **WP Sequence No.**

Made in the USA
Monee, IL
05 November 2026

67808357R00125